Anomalies and Scientific Theories

ANOMALIES
AND
SCIENTIFIC THEORIES

WILLARD C. HUMPHREYS

New College

 FREEMAN, COOPER & COMPANY
1736 Stockton Street
San Francisco 94133

Acknowledgments

Portions of this book are based upon the author's doctoral dissertation, "Anomalies and Scientific Theories," submitted to Yale University in June, 1966. The author is indebted to a number of friends for helpful comments and encouragement, especially the late Norwood Russell Hanson, Professors Henry Margenau and Frederick Fitch of Yale, J. Brookes Spencer of Oregon State University, David Hull of the University of Wisconsin-Milwaukee, Sidney Morgenbesser of Columbia University, and Rudolph H. Weingartner of San Francisco State College. For any errors of fact or judgment the author alone remains responsible. Acknowledgment is gratefully made to the following authors, journals, and publishers for permission to quote from copyrighted materials:

Cambridge University Press, New York (*Scientific Explanation*, R. B. Braithwaite).

Dover Publications Inc., New York (*Foundations of Nuclear Physics*, ed. R. T. Beyer).

Harper and Row, New York (*Causality and Chance in Modern Physics*, David Bohm; *The Philosophy of Science*, Stephen Toulmin).

Werner Heisenberg, "Uber den Bau den Atomkerne I" and "Uber den Bau den Atomkerne II," in *Zeitschrift für Physik*, vols. 77, 78, 88 (Berlin: Springer-Verlag).

Carl G. Hempel and Paul Oppenheim, "The Logic of Explanation," in *Readings in the Philosophy of Science,* eds. H. Feigl and M. Brodbeck (New York: Appleton-Century-Crofts).

The Macmillan Company, New York (*Aspects of Scientific Explanation,* Carl G. Hempel).

Methuen & Company, Ltd., London, and The Philosophical Library, New York (*Nuclear Physics,* Werner Heisenberg).

The Physical Review, American Institute of Physics, articles by S. H. Neddermeyer and C. D. Anderson in vols. 45, 50, 51, 85.

The Proceedings of the Royal Society A, London, vols. 159, 160, 161, 165, 174, 176. Detailed citations in footnotes.

Contents

Anomalies and Scientific Theories

Introduction

Science, like Janus, has two distinct aspects: an historical side and a logical or epistemological structure. It involves both a dynamic process of knowledge acquisition and the product of that process; viz., organized knowledge. The present study aims at achieving a deeper understanding of both the process and the product. It is a study in the history and philosophy of physical science.

On the historical side, we shall attempt to document one key phase in the development of the modern quantum theory. Chapter 6, a detailed account of events leading to the discovery of the meson, is entirely devoted to this subject. But earlier portions of the book (notably Chapter 1) are both logically and chronologically linked to it as well. Of necessity, the historical treatment is episodic. The history of optics, mechanics and electromagnetism since Newton is obviously beyond the scope of one slender volume. The episodes we shall emphasize, however, are deliberately chosen for their capacity to illuminate both the history *and* logic of modern physical theory. Newton's work on colors and light, Einstein's hypotheses on photoelectricity and the quantum of energy, and other subjects to be discussed all stand historically and logically behind the 20th century's conception of what physical explanation is and ought to be. We offer, then, not a comprehensive history of the quantum theory and its antecedents but, instead, a brief survey of some of the explanatory techniques of quantum and pre-quantum physics as evidenced in the actual history of these branches of science.

11

On the logical side, we shall attempt to evaluate critically some of the central and most widely accepted philosophical accounts of the nature of physical laws, explanations and theories. Chapter 2 (and Chapter 1, in part) emphasize and underline certain serious omissions in the well-known Deductive-Nomological model of scientific explanation. While accepting the main outlines of this model we shall try to show that it is radically incomplete and insufficient as a logical analysis of actual physical explanations.

In Chapter 3, the Hypothetico-Deductive account of scientific theories will be considered and, like the Deductive-Nomological account of explanation, found wanting. A modified and amended version of the H–D model will be sketched out—a version closer in spirit to the practice of working scientists and at the same time more faithful to the inferential patterns between theory and observation which actually obtain in quantum and pre-quantum mechanics.

Chapters 4 and 5, respectively, focus on the questions of how statistical laws function in theories like quantum mechanics and how the limits of explanation of a physical theory are to be determined. Both issues, of course, have been in the forefront of philosophical discussion of the quantum theory in recent years. They are moreover essential to a true historical understanding of the way in which the quantum theory has developed.

The central theme throughout will be the concept of "anomaly"—the concept of a fact or event which requires or demands explanation. *Our thesis, briefly put, is that both the logical structure of scientific theories and their historical evolution are organized around the identification, clarification and explanation of anomalies.* As an historical thesis this can only be defended by copious reference to the facts. As a logical thesis, however, several of its implications can be drawn out immediately by way of anticipation. For example:

(1) The conception of explanation as deriving its logical force from mere deduction of a description of the facts at hand from general laws is a drastic oversimplification. A physical explanation, we shall argue, depends implicitly on a context of assumptions against which the given fact is seen to require explanation. Contrary to the "deductivist thesis," therefore, it is not the deductive linkage which carries explanatory illumination but the *contrast* between the background context and the proffered explanatory statements. Even where the deductive connection is present it can fail to provide explanation unless the proper *logical* relations obtain between the contextual background, the description of the anomaly and the laws cited. This is not, we shall claim, a mere psychological or pragmatic point but one going to the very heart of the logic of explanation.

(2) Theories in the physical sciences, while having features in common with axiom systems of pure mathematics, will be shown to have special logical and semantic properties setting them apart. These properties arise as a result of the dual role played by theories in both *identifying* and *explaining* anomalies. To the extent that a theory is capable of locating a fact discordant with itself and then explaining that fact away the theory must not be a purely axiomatic system. If a purely axiomatic theory implies that a phenomenon p should not occur and p indeed does occur there will be no chance of explaining p via that theory. If anything, p must be counted as evidence against the theory. But as we shall subsequently show physical theories typi-

cally accomplish the task of identifying and explaining one and the same anomaly without being disconfirmed in any way.

In what follows, these points will be developed at greater length as the historical role played by anomalies in the shaping of physical theory comes under scrutiny.

1

Natural Anomalies

What makes a good scientific explanation?

The simplest answer to this question—though surely not the best—is that explaining an event consists in pointing out its cause. The word 'cause,' of course, is so ambiguous and vague that this is hardly very illuminating. But typically something fairly definite is meant when one speaks of causes in the context of scientific explanation. The idea is that some event or occurrence, antecedent to (or contemporary with) the one being explained, is to be singled out as the causal factor. Presumably, this event is singled out because of its "decisive" or "special" importance for the occurrence of the event we wish to explain. Thus, the cause of a match bursting into flames is generation of heat through friction in the striking of the flame. And the cause of an object rising on a block and tackle is the application of force at the other end of the chain.

It is extremely difficult to say what gives these particular antecedent events their "decisive" importance in the production of the effect events. In some sense, however, they are essential to the occurrence of the effect. As a first approximation we might say that they are "non-eliminable

15

conditions within a total set of conditions jointly sufficient for producing the event." [1] For there surely are a host of other relevant factors besides the heat which ignites the match and the force which raises the weight. Provisionally, therefore, we shall assume that a cause is simply a single condition selected from a longer list of conditions because of its special relevance in producing the effect. Its elimination from the list will result in the failure of the effect to follow.

The condition identified among all relevant conditions as "the cause" of the anomalous occurrence is not uniquely determined, of course. Many factors included in a full list of the jointly sufficient set of conditions may be "essential" in the indicated sense. Thus,

---

[1] Some writers regard the cause of an event as a *sufficient condition* for its production. Thus, Cohen and Nagel say: "By the *cause* of some *effect* we shall understand, therefore, some appropriate factor invariably related to the effect. If *A has diphtheria at time t* is an effect, we shall understand by its cause a certain change *C*, such that the following holds. If *C* takes place, then *A* will have diphtheria at time *t*; and this is true for all values of *A*, *C*, and *t*, where *A* is an individual of a certain type, *C* an event of a certain type, and *t* the time." Morris R. Cohen and Ernest Nagel, *An Introduction to Logic and Scientific Method* (New York: Harcourt, Brace and Co., 1934), p. 248. The only real difference between this account of causality and the one we have discussed is that it ignores the contextual conditions accompanying *C* or assumes that mention of them has been included in *C*. See also, John Hospers, *An Introduction to Philosophical Analysis* (New York: Prentice-Hall, Inc., 1953), pp. 242–45.

The cause of an outbreak of plague may be regarded by the bacteriologist as the microbe he finds in the blood of the victims, by the entomologist as the microbe-carrying fleas that spread the disease, by the epidemiologist as the rats that escaped from the ship and brought the infection into the port.[2]

Context and intellectual interest lead us to select one possible cause here rather than another. For this reason, many logicians favor viewing 'the explanation' and 'the cause' as terms referring to the *total* set of antecedent conditions rather than any one part of it.[3] When this way of speaking is adopted, the answer to the original question is modified slightly: giving a good scientific explanation consists in pointing to the totality of relevant empirical conditions leading to the anomalous event. *The* explanation of the plague, then, includes reference to all of the factors—rats, fleas and microbes—rather than to one alone.

A somewhat more sophisticated analysis of scientific explanation will be considered below.[4] For the time being let us assume that the foregoing analysis in terms of a jointly sufficient set of conditions is a reasonable general description of the norms operative in actual scientific prac-

---

[2] W. I. B. Beveridge, *The Art of Scientific Investigation* (New York: Vintage Books, 1957 (1950)), p. 126.

[3] Hospers, *op. cit.,* p. 244: ". . . we often talk as if a *part* of the sufficient condition were the cause, though we realize well enough that until we have stated the entire sufficient condition we have not stated the *whole* cause."

[4] See Chapter 2.

tice. What scientists seek, then, is knowledge of the totality of relevant conditions of an event. What they can be expected to reject is any alleged explanation which does not implicitly or explicitly lay out the total set of conditions.

The implications of this elementary analysis of the logic of explanation are striking. Suppose, for instance, that we couple it with the traditional metaphysical dictum that every event has a cause (though the state of human knowledge at any given time may not reveal to us what that cause is). *From this set of premises it follows that, in principle, all events are equally capable of being explained.* No intrinsic differences among events mark some off as being "more explicable" than others.

The situation can be nicely summed up by means of a simple metaphor. Nature presents us with an unbroken string of identical pearls: the facts. The logic of explaining one is the same as the logic of explaining any other. One merely lists all of the relevant antecedent conditions (a completely empirical matter) and has done with it. Nor is there any distinction of an observable nature between events which would enable us to sort them out as being more or less explainable. Identical pearls all look alike. And events do not come with little tags listing their degree of abnormality. Any differences in "explainability" must therefore stem from extrinsic sources, e.g., practical considerations, aesthetic affections or human desires and interests.

On the basis of these considerations, pragmatism—or a version of it—seems to follow. Our basis for discriminating among events as regards explainability must be a matter of non-logical and non-empirical considerations. One pearl catches our eye and captures our attention, but there is no rationale for this beyond our interests, affections and feelings. The scientist who chooses to explain an event, then, merely invokes his interests as the criterion of choice. It is, as it were, a matter of taste which event he selects. On such fancies depends the direction of scientific research.

Against this strange view, common sense suggests that some events are more in need of explanation than others, and not merely because of personal or social preferences. When we say, for instance, that the great Alexandrian astronomer, Claudius Ptolemy, was obliged to explain the retrograde motions of Mars or the proximity of Mercury to the sun we are not expressing an arbitrary preference of our own. Instead, we are pointing to an intellectual and conceptual requirement built into Ptolemy's own beliefs and cognitive commitments. If Ptolemy had failed to seek an account of these phenomena, yet persisted in his general beliefs about the nature of planetary motions, we should be forced to conclude that he had at least misconceived the nature of the enterprise of theoretical astronomy. There would be no use in Ptolemy's replying to our demand for an explanation by saying "Your goals and interests and purposes lead you to seek an explanation of these facts but my inclinations lead me elsewhere." Given his commitments regarding the nature of the planetary motions this answer is logically excluded. And, in general, given the human race's elementary common sense beliefs about nature there will always be some facts or events whose existence cries out for explanation.

In one respect, the conception of facts as a string of identical pearls contains a kernel of truth. *For it is certainly possible to construe any given fact as an anomaly.* All one need do is to put oneself in the right frame of mind and it becomes possible to see the presence of a nose on one's face as one of the queerest occurrences imaginable. The mental process is very much the same as that involved in so-called gestalt shifts. Now the event or fact is "seen as" a normal state of affairs; now it is "seen as" requiring explanation. Now the psychologist's picture appears as a duck, now as a rabbit.[5] It is all a matter of one's way of looking at things.

But the fact that any event can be "seen as" requiring explanation does not warrant the further assertion that all events or facts stand equally in need of explanation. Some modes of "seeing as" are forced upon us—logically speaking—while others are merely contrived.[6] Some events or facts are anomalous only as a result of mental gymnastics on our part. Others are "naturally anomalous," *vide* Ptolemy's problems with Mercury and Mars.

What needs clarification here is the idea of *natural anomalousness;* the idea of an event being, in some sense, *objectively* in need of explanation. In order to get clear about this notion let us begin by considering in their historical context several anomalous states of affairs which have played decisive roles at one point or another in the development of modern physical concepts. In Chapter 2 we shall then seek to round out our treatment of natural

---

[5] Ludwig Wittgenstein, *Philosophical Investigations,* tr. by G. E. M. Anscombe (New York: The Macmillan Co., 1953), p. 194.

[6] N. R. Hanson, *Patterns of Discovery* (Cambridge: Cambridge University Press, 1961), p. 15: ". . . I must talk and gesture around . . . to get you to see the duck when only the rabbit has revealed itself. I must provide a context. The context is part of the illustration itself. Such a context, however, need not be set out explicitly. Often it is 'built into' thinking, imagining and picturing. We are set to appreciate the visual aspect of things in certain ways."

anomalies by providing a detailed logical analysis. Our aim is twofold: on the one hand, to understand the ways in which anomalies function in the historical process of scientific inquiry; on the other hand, to understand the logical and formal conditions which are involved in making an event or fact stand out as anomalous.

Hypotheses Non Fingo

Isaac Newton's first published paper, an exposition of certain experimental results on light and the theory of colors, appeared in 1672.[7] The furor it touched off could hardly have been anticipated,[8] and from the standpoint

---

[7] Newton's paper is reproduced in *Isaac Newton's Papers and Letters on Natural Philosophy*, ed. by I. B. Cohen (Cambridge: Harvard University Press, 1958). See especially pp. 47–51. Cited hereafter as *Newton's Papers*.

[8] A contrary view is expressed by Sir Edmund Whittaker in his *History of the Theories of Aether and Electricity* (New York: Harper, 1960), Vol. I, p. 18: "The publication of the new theory gave rise to an acute controversy. As might be expected, Hooke was foremost among the opponents, and led the attack with some degree of asperity." Whittaker's opinion, however, is based on the assumption that on account of Newton's work "Hooke's theory of colour was completely overthrown . . ." (*Ibid.*, p. 17) As will become clearer momentarily, this is a misleading interpretation of the significance of Newton's results. Hooke believed—and Newton did not dispute the point seriously—that his hypotheses on color were compatible with Newton's main experimental results. As Richard Westfall puts it, "Newton told Hooke how Hooke's hypothesis could be reconciled with the newly discovered property of light and

of the modern observer is almost incomprehensible. Young Mr. Newton quickly found himself engaged in a wide-ranging debate on optics, light and scientific method with some of the world's leading scientists. In fact, the heat generated in the youthful upstart's exchanges with Robert Hooke of the Royal Society led to permanent hard feelings between the two.

What did Newton say to bring on this holocaust? It is difficult at first to see. His paper seems merely to report on and elaborate direct accounts of observation and experiment. Little in the way of "theory" is presented and this is carefully cloaked in the terminology of supposition and guarded assumption. It is hard to imagine a less controversial or polemical approach. And yet the kernel of controversy is clearly there. It appears, as we shall shortly show, in Newton's discovery of an anomaly which his contemporaries were unable to perceive as such.

Newton's paper divides fairly neatly into three parts. The first is the "straight research report" on work he had done on refraction of light through glass prisms. The second portion attempts to show the implications of this work for the construction of refracting telescopes. (Only a short time before submitting his paper for publication Newton had given the Royal Society a new *reflecting* telescope of his own invention. It is clear that this second part of the paper is intended to explain his reasons for having attempted a new type of construction.) At any rate, there is

---

ended by calling the hypothesis 'insufficient' and 'unintelligible' . . ." This hardly constitutes the "overthrow" of Hooke's theory. Cf. Richard Westfall, "Newton's Reply to Hooke and the Theory of Colors," *Isis*, Vol. 54 (1963), p. 88.

to be found here an argument for the view that refracting telescopes—by virtue of their refraction of light—are inherently subject to a certain amount of chromatic aberration and hence subject to ultimate limitations of usefulness.

Finally, in the third portion of the paper we find Newton returning to the experimental report. He lists several conclusions concerning the colors produced by prismatic refraction of white light, interspersing a few guarded conjectures on the possible implications of these results for the theory of the physical structure of light. Newton thought his results lent some degree of credibility to the hypothesis that light is corpuscular rather than wave-like.

The first part of Newton's paper, the part which claims our closest attention here, gives an account of his solution of a perplexing problem about the prism. According to his report, he was toying with a prism and a circular beam of sunlight let in through a hole in a shutter when a peculiar anomaly caught his attention. As was well-known to Newton's contemporaries, the refraction of such a beam through a prism produces a rainbow-hued spectrum of colored light. Newton was investigating this phenomenon with an eye to the improvement of refracting telescopes (whose tendency to produce color fringes was equally well-known). Here are Newton's own words as he describes the spectrum of colors thrown on the far wall: ". . . I became surprised to see them in an *oblong* form; which, according to the received laws of Refraction I expected should have been circular." [9] The "received laws" here cited, of course, are simply the principle of rectilinear propagation of light and

---

[9] *Newton's Papers,* p. 48.

the then current version of Snell's law—"Cartes' (Descartes') hypothesis" Newton calls it.[10]

What Newton means in this passage is that the rays of sunlight in the circular beam ought all to be refracted equally by the prism in accordance with Descartes' principle. They should therefore remain parallel to one another all the way to the wall. Assuming that this occurs, the image of the spectrum should have a shape geometrically similar to the shape of the hole admitting the light. That this does not occur implies clearly the presence of some defect either in the system of "received laws" or in the experimental setup. The elongation, in a word, is anomalous.

Newton passes on to us several hypotheses he entertained and tested experimentally in order to arrive at an explanation of the spreading of the spectrum. He considered, for instance, the possibility that the refracted light might have a non-rectilinear path, i.e., that some rays on the edge of the beam might be induced to travel along a curved trajectory. This proved not to be the case. He also went on to examine systematically several relevant variables such as the texture and homogeneity of the glass, the position of the prism with respect to the hole, the apparent angular width of the sun's disc and the position of the prism with respect to the dark region inside the room and the sunny region outside. None of these factors provided an explanation of the

---

[10] Isaac Newton, *Optical Lectures Read in the Publick Schools of the University of Cambridge Anno Domini 1669* (London: Printed for Francis Fayram at the South Entrance of the Royal Exchange, 1728), pp. 13, 50, for example. The main result of Newton's 1672 paper is embodied in these lectures but they were not immediately published.

spreading effect. Rays drawn from opposite sides of the sun's disc would not diverge sufficiently far on refraction to produce the observed effect. No significant inhomogeneity could be detected in the glass. And so on through the list. There was no easy explanation of the anomaly.

At last Newton hit upon an experiment—he called it the *experimentum crucis*—which resolved the issue. In this experiment, clusters of individual rays of color-bearing light are isolated by means of two baffles or screens. (*See Diagram 1*). The monochromatic light beam thus isolated is then refracted through a second prism at *B*. By rotating Prism *A* slightly, light from different parts of the spectrum cast on Baffle 2 can be supplied to Prism *B*. The result is that the spot of light cast on the wall simultaneously changes color and moves up and down the wall *without*

Diagram 1

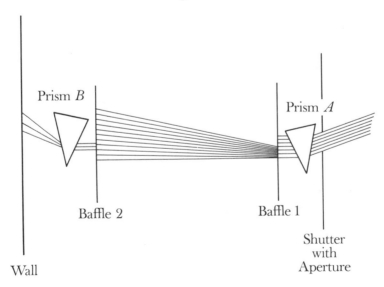

Prism *B*

Prism *A*

Baffle 2

Baffle 1

Shutter
with
Aperture

Wall

Prism *B* being moved at all. The evidence is wholly un-ambiguous: different portions of the beam emerging from Prism *A* are refracted differently at Prism *B*.

And so the true cause of the length of the Image was detected to be no other, then that *Light* consists of Rays differently refrangible, which, without respect to a difference in their incidence, were, according to their degrees of refrangibility, transmitted towards divers parts of the wall.[11]

As a corollary to this major conclusion, Newton argues that the different colors produced in the spectrum are each associated with light of a particular degree of refrangibility. In short, all of the different color-bearing rays are found to be components of the original beam of white light. Each is distinguishable by its degree of refrangibility.

All of these results are so straightforwardly suggested to us by the facts that the response of Newton's scientific elders is difficult to comprehend. Aside from the suggestion by Lucas that Newton had botched the *experimentum crucis*,[12] there was the more general feeling that he had proved nothing worth bothering the Royal Society about. Hooke, taking his cue from Newton's brief references to

---

[11] *Newton's Papers,* p. 51.

[12] Lucas' claim is not entirely absurd. In his repetition of the experiment he found an elongation of the spectrum not five times the width but about three and one-half times. The dis-crepancy appears to have been due to differences in the dis-persive power of the kinds of glass being used. See Florian Cajori, *A History of Physics* (New York: Dover, 1962), pp. 92–93.

the corpuscular theory of light in the third portion of the
paper, asserted flatly that Newton's problem about the
spectrum could just as well be solved by several other
hypotheses, including a ramified wave theory.[13] It is pain-
fully clear, however, that Hooke did not have the vaguest
idea what Newton's problem really was. He evidently believed that Newton was putting forward a hypothesis to
explain the entire physical mechanism of color production
inside the beam of light. Thus, he concluded his first
rejoinder against Newton in the following way:

> If Mr. Newton hath any argument that he supposeth
> an absolute Demonstration of his theory, I should be
> very glad to be convinced by it: the Phaenomenon of
> light and colours being, in my opinion, as well worthy
> of contemplation as anything else in the world.[14]

The reference here to "absolute Demonstration" presumably refers to Newton's use of the expression '*experimentum crucis.*' By this phrase Newton meant merely an important or noteworthy experiment. Hooke, in keeping with
the more modern usage of the expression 'crucial experiment,' seems to assume that Newton's experiment was intended to decide once and for all between the corpuscular
hypothesis and all other alternatives.[15] Looked at in this

---

[13] H. W. Turnbull, editor, *The Correspondence of Isaac Newton* (Cambridge: Cambridge University Press, 1959–), Vol. 1,
pp. 110–14.

[14] *Ibid.,* p. 114.

[15] In a second letter Hooke complains: "All that he doth prove
by his *Experimentum Crucis* is that the colour'd Radiations

way, it is no wonder Hooke found the Newton paper unsatisfactory. Newton's *experimentum crucis* hardly counts
as a disproof of the wave theory of light. It was not intended
to.[16] It is designed merely to identify the cause of the
anomalous elongation.

Newton's subsequent reply to Hooke's criticism is scath-
ing. There is no question about it. He had been badly

---

doe incline to ye Ray of light with Divers angles and that they
doe persevere to be afterwards by succeeding mediums diversly
refracted one from another in the same proportion as at first,
all wch may be, and yet noe colour'd ray in the light before
refraction . . ." (Newton's Correspondence, *op. cit.*, Vol. 1,
p. 202.) Hooke is clearly construing Newton's remarks about
different rays producing different colors as though Newton
were speaking of streams of light particles rather than geometrical abstractions. From this standpoint he is quite right in
claiming that Newton has not shown the existence of differently
colored rays (streams of particles) in the original light beam.
Newton's result is quite compatible with a wave theory of light.
[16] In his *Optical Lectures* (London: Printed for Francis Fayram
at the South Entrance of the Royal Exchange, 1728), however,
Newton says: ". . . since I observe the Geometers hitherto
mistaken in a particular property of Light, that belongs to its
Refractions, tacitly founding their Demonstrations on a certain
Physical Hypothesis not well established; I judge it will not be
unacceptable if I bring the Principles of this Science to a more
strict Examination, and subjoin, what I have discovered in
these Matters, and found to be true by manifold Experience,
to what my reverend Predecessor [Isaac Barrow] has last deliv-
ered from this Place." *Loc. cit.*, p. 2. The "Physical Hypothesis" here mentioned is probably "Cartes' Hypothesis," i.e.,
Snell's law. It is possible, however, that Newton is administer-
ing a slap at the whole Cartesian theory of light.

hurt by the famous scientist's failure to understand his paper. Seizing on Hooke's references to "hypotheses" he boldly denied that hypotheses have any role in natural philosophy.[17] Hooke was shocked:

. . . for I judge there is noething conduces soe much to the advancement of Philosophy as the examining of hypotheses by experiments & the enquiry into Experiments by hypotheses.[18]

Newton was not to be put off by this philosophical truism. He continued to insist that his opponent's error consisted in the "logical" blunder of hypothesizing instead of "deducing from the phenomena." With the exception of Ernst Mach, who claims that Newton "explained facts, while Hooke wanted to wrangle over hypotheses," [19] the

---

[17] Newton's Correspondence, op. cit., Vol. 1, p. 177. E. A. Burtt, in his Metaphysical Foundations of Modern Science (Garden City, N. Y.: Doubleday Anchor Books, 1932), p. 223, suggests that Newton would not have objected to hypotheses as guides to experiment during his early days of optical work. This does not square well with Newton's expressed views during the period and can only be maintained if we allow a considerable latitude in the application of the term 'hypotheses.' See also Newton's Papers, pp. 93–94.

[18] Newton's Correspondence, op. cit., Vol. 1, p. 202.

[19] Ernst Mach, The Principles of Physical Optics, tr. by J. S. Anderson and A. F. A. Young (New York: Dover Publications, Inc., reprint of 1926 edition), p. 87. Mach goes on to say: "His [Newton's] conflict with misconception and the inexact repetition of his experiments by Pardies, Linus, and others, must have depressed him. Thus it is quite conceivable that there

consensus of recent philosophers and historians has been that Newton erred in attacking the employment of hypotheses. William Kneale puts it this way:

> Newton's doctrine . . . of scientific method is very puzzling, because it does not square with his own practice. He was obviously inclined to favour the atomic hypothesis about matter and the corpuscular theory of light, and he allowed himself occasionally (e.g. at the end of his *Optics*) to make speculations of a transcendent character about the explanation of gravitation. But more remarkable still, his establishment of the theory of motion and of the principle of universal gravitation, which he cites as an example of direct induction from phenomena, is in truth a very notable achievement of the hypothetical method.[20]

At any rate, Newton soon found himself deeply enmeshed in a methodological debate in which he is cast as the experimentalist heavy while his adversaries appear as enlightened champions of the modern theoretical approach and hypothetico-deductive method.[21]

---

probably grew up a certain amount of insincerity and restraint in the exposition of Newton's researches. In no other place did he expound his method of investigation so clearly and openly as he did in his earliest optical publication."

[20] William Kneale, "Induction, Explanation and Transcendent Hypotheses," in *Readings in the Philosophy of Science,* Herbert Feigl and May Brodbeck, eds. (New York: Appleton, Century, Crofts, Inc., 1953), p. 358.

[21] A full discussion of the hypothetico-deductive method is given below in Chapter 3.

Newton's own understanding of the logic of the dispute with Hooke is surely inadequate. Yet he was right in believing that Hooke and others were confused on a key point of methodological principle. The point is that the fact or phenomenon which really required explanation was not, as Hooke seems to have believed, the mere presence of colors on the wall. Nor was it the whole complex of facts involved in the linkage between prismatic refraction and color production. There was but one *natural anomaly* in the circumstances and this was the length of the spectrum. That fact and that fact alone had to be put into proper perspective and explained.

There is a touch of subtle irony in Newton's defense of his own work against the criticisms of Hooke. The irony is generated by the contrast between Newton's professed methodology[22]—*"hypotheses non fingo"*—and his *de facto*

---

[22] For other formulations of Newton's methodology, see (a) *Newton's Papers,* p. 93: ". . . the proper Method for *inquiring* after the properties of things is to deduce them from Experiments." (b) Newton's *Opticks* (New York: Dover Publications, Inc., 1952), p. 369: ". . . the main Business of natural Philosophy is to argue from Phaenomena without feigning Hypotheses, and to deduce Causes from Effects . . ." (c) *Newton's Principia,* Cajori's revision of the Motte translation (Berkeley: University of California Press, 1960), p. 547: ". . . I frame no hypotheses; for whatever is not deduced from the phenomena is to be called an hypothesis; and hypotheses, whether metaphysical or physical, whether of occult qualities or mechanical, have no place in experimental philosophy. In this philosophy, particular propositions are inferred from the phenomena, and afterwards rendered general by induction. Thus it was that the impenetrability, the mobility, and the impulsive force of bodies, and the laws of motion and of gravitation, were discovered."

awareness of the delicate balance and interplay between theory and observation, hypothesis and experiment. Indeed, it is Newton's obvious sensitivity to the importance of theory as a guide to observation which sets him apart from his critics. Hooke was oblivious to the elongation of the spectrum. From his standpoint it represented just one more fact to be explained by a general account of the refraction process—just one more pearl on the string, to recall our earlier metaphor. For this reason, Hooke viewed Newton's corpuscular hypothesis as the core of the paper. Clearly, any general account of the whole cluster of facts concerning the spectrum would start from some such premise.

Newton, on the other hand, recognized the clear and present danger to Snell's law and the whole theory of geometrical optics posed by the elongated spectrum. In the absence of a *geometrical* solution of the problem, all hypothesizing about physical optics loses its point. One can hardly develop photon, wave or particle theories until it is settled whether, for instance, the path of light rays emerging from the prism is rectilinear or curvilinear. Accordingly, it is Newton—not the hypothesizing Hooke —who leads the way toward the development of modern optical theory. Hooke, with his passion to explain everything at once, leads us only to one of the many conceptual dead ends in the history of science.[23]

The distinguishing mark of Newton's approach is his penetrating grasp of the naturally anomalous character

---

[23] "Hooke tried to explain colour, but it was Newton who first correlated it with differing periodicities." (H. T. Pledge, *Science Since 1500* (New York: Harper, 1959), p. 69.)

of the spectrum elongation. His careful experimentation to verify the existence of an anomaly is perhaps even more important. Thus, rather than attempting the nebulous and elusive task of "explaining the phenomena of light and colors" (whatever that might mean!), Newton tackles the well-defined problem at hand. And in this way he succeeds in providing more insight into the nature of light than a thousand of Hooke's hypotheses would ever provide.

The important philosophical moral in all of this is signaled in the fact that Newton perceives the extraordinary character of the elongated spectrum *because* of his theoretical beliefs. What seems obviously to be implied is that natural anomalies are what they are on account of their logical status *vis-à-vis* existing theory, the "received laws," as Newton put it. Their anomalous character derives directly from their being in conflict with accepted beliefs.

The fact that Newton was able to *prove* that the elongation was anomalous—to his own satisfaction and the modern reader's if not to Hooke's—is likewise important. For it shows that the aura of peculiarity is not a mere personal fancy but an objective, logical property of the anomaly. There are, in a word, rational criteria for identifying anomalies.

Finally, we need to note that the role of theory as the source of Newton's insight into the importance of the elongation says something of significance about the nature of physical theory itself. *Apparently it is a fundamental function of physical theories to single out certain events and facts as requiring explanation.* Newton's success (and Hooke's failure) in employing the principles of Cartesian optics to this end shows precisely how fundamental the function is. Theory serves him as a probing instrument with which to order observation and experiment. In the

end it provides explanation. But in the beginning it speci-
fies what is going to need explaining.

But perhaps we are overstating the case. Perhaps it is a
peculiarity of Newton's particular technique, or even a
peculiarity of geometrical optics itself, which leads us to
assign such a role to theory. Let us turn to an entirely
different realm of physics, a different period of history and
the work of a very different kind of physicist in order to
test the validity of our generalization.

NEPTUNE AND MERCURY

U. J. J. Leverrier's role in the discovery of Neptune and
in the subsequent failure to explain the behavior of the
planet Mercury has been aptly characterized as spanning
the "zenith and nadir" of Newtonian celestial mechanics.[24]
Leverrier's theoretical prediction of the existence of Neptune signals the high water mark of the theory's predictive
power. And his inability to extend the techniques used
there to the problem of explaining the advance in the
perihelion of Mercury signals the advent of the relativity
era. But his work also stands as the culmination of the
development of explanatory technique within Newtonian
science and it is partly for this reason that we turn to it
here. Leverrier's decisive identification of natural anomalies in the behavior of Mercury and Uranus—his explanation of the latter and his failure to explain the former—
are the factors which explicitly define the logical boundaries of explanation within classical mechanics.

---

[24] N. R. Hanson, "Leverrier: The Zenith and Nadir of Classical
Mechanics," *Isis*, Vol. 53 (1962), pp. 359–78.

At least as early as 1821, when Alexis Bouvard sought to tabulate the elements of Uranus' orbit via the techniques of Laplace's *Mécanique Céleste,* it was evident that something was amiss in the planet's behavior. Eighteenth century observations, though few in number, had to be dropped in order to perform the necessary calculations.[25] This, even though the integrity of the early reports was not seriously to be doubted.

By 1840 the failure of any competent astronomer to attack the problem seriously had become something of a celestial scandal. Uranus' observed positions were daily becoming more distant from the positions generable from Bouvard's tables. It was at this point that Leverrier, urged on by F. Arago, entered the picture. (Almost simultaneously, but independently, the problem was taken up by J. C. Adams in England. We shall not here go into the tragic story of Adams' failure to get a public hearing and full credit for his work.)[26]

Leverrier's efforts resulted in the publication of three important papers during the years 1845 and 1846. In the first of these,[27] he reports his initial attempt to correct the

---

[25] U. J. J. Leverrier, "Premier Mémoire sur la Théorie d'Uranus," *Comptes Rendus hebdomadaires des séances de l'Academie des Sciences* (cited hereafter as *Comptes Rendus*), Vol. 21 (1845), p. 1051.

[26] For Adams' own account of the matter, see *A Source Book in Astronomy,* ed. Harlow Shapley and Helen E. Howarth (New York: McGraw-Hill, 1929), pp. 245–48—a reprint of a paper by Adams which appeared in the *Memoirs* of the Royal Astronomical Society in 1847.

[27] "Premier Mémoire . . . ," *op. cit.*

Bouvard tables for the perturbing influences of Jupiter and Saturn. His calculations purport to show that for 1845, even the adjusted Bouvard tables leave an unexplained lag of 40 seconds in the predicted longitude of the planet.

The first paper, then, sets out a clearly defined anomaly requiring explanation. Soon afterward, however, Leverrier discovered that the anomaly was not so well-defined as he had thought.[28] For a number of vitiating sources of possible error—none of which could be estimated with sufficient accuracy—had gone into the construction of Bouvard's tables to begin with. Thus, the mere tinkering he had done with the table was not sufficient to show conclusively the existence of a 40 second gap between predicted and actual position. The entire table would have to be recalculated from scratch, i.e., from the previously recorded observations.

The results of this tedious and difficult labor are announced in the second paper (June, 1846).[29] After having taken into account all previous observations whose credentials were sound Leverrier was able to conclude that the planet was indeed out of position. "The observations and the theory are irreconcilable." [30] Thus, the existence of an anomalous state of affairs had been established beyond all reasonable doubt. Once again it is theory which provides the conceptual background against which the anomaly is identified and defined. *Like Newton, Leverrier demonstrates the need for an explanation by appeal to*

---

[28] U. J. J. Leverrier, "Recherches sur les mouvements d'Uranus," *Comptes Rendus,* Vol. 22 (1846), p. 909.

[29] *Ibid.*

[30] Hanson, *op. cit.,* p. 361 and Leverrier, *ibid.,* pp. 907–18.

existing theory. This time, however, it is the theory of classical mechanics, not geometrical optics, which provides the intellectual setting. And it is a purely observational (as opposed to experimental) anomaly which is defined. With Leverrier's demonstration of an anomaly, the way was opened for consideration of explanatory hypotheses. As a matter of fact, some astronomers had not even waited for Leverrier's proof before advancing conjectures. A possible satellite of Uranus had been suggested as the cause of the trouble and some of the more adventurous were even hinting that perhaps Newton's law of gravitation does not hold over very great distances.[31] To Leverrier, the drawbacks of these alternatives were obvious. A sound theory should not be abandoned simply because anomalies have arisen. An explanation *within* the theory's logical limits may still be possible. He therefore chose to follow up a third, more plausible hypothesis: the hypothesis that an as yet unobserved planet beyond Uranus was perturbing its motion. To test this hunch, he carried out difficult inverse-perturbation calculations of the probable orbit and size of such a planet. He was obliged at every step of the way to square his assumptions about the planet with the patent fact that no astronomers had ever observed it.

Some of his preliminary results were announced in the

---

[31] Leverrier, "Premier Mémoire . . . ," *op. cit.,* p. 1050: ". . . chaque jour Uranus s'écarte de plus en plus de la route qui lui est tracée dans la Éphemérides. Cette discordance préoccupe vivement les astronomes, qui ne sont pas habitués à de pareils mécomptes. Déjà elle a donné lieu à un grand nombre d'hypotheses. On est même allé jusqu'à mettre en doute que le mouvement d'Uranus fût Rigoureusement soumis au grand principe de la gravitation universelle." And so forth.

second paper already alluded to. But the full details emerge
only in the third memoir (August 31, 1846).[32] Here Leverrier gives a definite prediction as to the region of the sky
in which the new planet can be observed and an estimate
(3″) of its apparent diameter. This information Leverrier
also forwarded directly to a number of observatories with
requests that the planet be sought out in the sky.

> On Sept. 23, 1846, Dr. Galle (a Berlin astronomer)
> received Leverrier's letter asking him to seek the planet.
> Within an hour Galle and his assistant D'Arrest no-
> ticed a star (of the eighth magnitude) unrecorded on
> Bremiker's celestial map. Next day he and D'Arrest
> observed the star in another place. Here was Leverrier's
> planet! [33]

---

[32] U. J. J. Leverrier, "Sur la planète qui produit les anomalies
observés dans le mouvement d'Uranus," *Comptes Rendus,* Vol.
23 (1846), pp. 428–38, 657–62.

[33] Hanson, *op. cit.,* p. 363. See also Leverrier's follow-up paper
"Comparison des Observations de la nouvelle planète avec la
théorie deduit des perturbations d'Uranus," *Comptes Rendus,*
Vol. 23 (1846), p. 741. Galle's discovery had nearly been anticipated by J. Challis, Director of the Cambridge Observatory.
After reading of Leverrier's calculation, ". . . which agreed
closely with that of Adams, Challis began to make a search
in July and August by registering at different days all the stars
in a field about the assigned place, to see whether one among
them changed its position. Diverted by other work, he then
failed to reduce and compare his observations; otherwise he
would most certainly have discovered the planet, for it was
among the registered stars." (A. Pannekoek, *A History of
Astronomy* (New York: Interscience, 1961), p. 360.)

The planet, of course, was Neptune. And the response from the scientific community at large was overwhelming. Leverrier was told that his name would "be linked forever with the clearest proof of universal attraction one can imagine." [34]

In point of fact, Leverrier's name today is linked more emphatically with one of the clearest *disproofs* of the theory of universal gravitation! For when he later sought to extend the method which had so successfully resolved the anomaly in Uranus' behavior to the stubborn contrariness of Mercury the result was the ill-fated hypothesis of the "planet" Vulcan.[35] This mythical object, persistently "sighted" by observers down to the turn of the century and beyond, was postulated by Leverrier as a body lying between the sun and Mercury. Its non-existence—plus the failure of all other hypotheses[36] about possible perturbing influences—is one of the strongest factual bases for present day acceptance of a relativistic rather than Newtonian theory of mechanics.

Leverrier's spectacular success and equally spectacular failure are less germane to our present interest than the logical foundations upon which both are based. For his

---

[34] See Encke's letter, *Comptes Rendus,* Vol. 23 (1846), p. 660.

[35] See Leverrier's paper in *Comptes Rendus,* Vol. 49 (1859), p. 381, and Lescarbault's letter, *Comptes Rendus,* Vol. 50 (1860), p. 40.

[36] "Different explanations [of the behavior of Mercury] were suggested—an undiscovered planet inside the orbit of Mercury; diffused attracting matter about the sun; a small variation in Newton's law, consisting in an increase in the exponent 2 by 1/6,000,000. None of them was satisfactory." (A. Pannekoek, *op. cit.,* p. 363.)

success and failure are as much—if not more—a matter of his careful identification of the anomalies in Uranus' and Mercury's orbits as a matter of his having proposed bold, new hypotheses. To put it bluntly: any fool can dream about hidden planets and undiscovered causes. Only a brilliant theoretician can tie such dreams down to the facts. The distinction is that between idle metaphysics and genuine science.

Let us be clear about the nature of Leverrier's undertaking: he does not set out with some nebulous program of "explaining the universe," or even "explaining Uranus." Instead, he spends considerable effort just deciding what it is that needs to be explained! Explicit definition of the anomaly appears here as a logical prerequisite to explanation. The entire first memoir and much of the second are given over to it.

The hypothesis, when finally formulated, is not advanced as a "mere" speculation. This was the error of those predecessors of Adams and Leverrier who had conjectured limply about satellites and hidden masses. In lieu of such vagueness Leverrier offers a detailed elaboration of the hypothesis: an account designed to explain the anomaly in full should the hypothesis be true. At the same time, he sets the stage for a decisive empirical test of the theory of Newtonian mechanics. For in the course of analyzing the anomaly it is necessary to set out data which sharply delimit the range of hypotheses compatible with Newtonian theory yet capable of explaining the facts. The hypothesis of a small, hard to observe satellite perturbing Uranus, for instance, is not capable of explaining the data unless one is prepared to make all manner of implausible assumptions about the eyesight of past observers, the density of the satellite and so forth.

True, Newtonian mechanics does not come crashing to the ground with the discovery of the advance of the perihelion of Mercury. But this is a practical, not a logical, matter. For it remains logically possible that some of the implausible assumptions needed to save the theory may yet be true. There is no *a priori* guarantee against a lead satellite with solid uranium core, for example. In time, though, further observation and further theoretical investigation (of, for instance, the conditions required for a stable orbit of a uranium-lead satellite) can settle such issues. The important point is that there is an upper bound on the number of such questions. As Leverrier observes, there were in actuality only three sources of error in the positional predictions for Uranus:

. . . (1) the error involved in the recent observation (which is) compared with the calculated elements of the orbit; (2) the uncertainty which can affect the calculated position as a result of errors in the longitudes which served as the basis for determining the elliptical elements; (3) finally the theoretical error due to some unknown secondary force which the planet actually is affected by.[37]

His own work systematically ruled out all of the (finitely many) items falling under (1) and (2). And, though greater

---

[37] Leverrier, "Recherches . . . ," *op. cit.*, p. 911. "Elles sont au nombre de trois, savoir: 1º l'erreur propre de la nouvelle observation comparée; 2º l'incertitude qui peut affecter la position calculée, par suite des erreurs des longitudes qui ont servie de base à la détérmination des éléments elliptiques; 3º enfin l'erreur théorique due à ce que la planète obéirait réellement a

effort is involved, enumeration and consideration of the possibilities under heading (3) is clearly feasible.

Leverrier recognized—and his more speculative predecessors seem to have overlooked the fact—that the logical resources of a well-made physical theory are not easily exhausted. (This is not the same as saying they are inexhaustible, of course.) In some cases the theory can *both* identify and explain an individual anomaly. It need not collapse in the face of the first apparent counterevidence to come along but may well succeed in bringing the anomalies it encounters under the umbrella of explanation. Only by carefully defining the anomaly and demonstrating its status with respect to theory can we tell for sure. Newton performed this task for geometrical optics and the elongated spectrum. Leverrier did it for classical celestial mechanics, Uranus and Mercury.

Once again the role of theory in fixing what requires explanation emerges as central and fundamental. Leverrier's problems are wholly specified for him by classical theory and his exploration of them is carried on wholly within the conceptual framework of that theory. Without it, the observational data on Uranus and Mercury which set his thinking in motion would have had no particular significance: they would simply have been so many pearls on the string. We can only speculate whether in that case the planet Neptune would ever have been discovered in the 19th century.

---

quelque force secondaire inconnue. Si nous prouvons que les deux premières causes ne sauraient suffire pour expliquer la différence qui existe entre le calcul et l'observation, nous serons forcés d'admettre l'influence de la troisième."

THE PHOTOELECTRIC EFFECT

Newton's use of theory in the identification of anomalous data was not an isolated case. The same kind of use occurs in Leverrier's work on Uranus and again on Mercury. Just how pervasive the pattern is we shall now try to show by considering the culmination and collapse of another physical theory: Maxwell's electromagnetic theory of light. Here again the guiding role of theory in the identification of natural anomalies will be manifest. But there will be certain important differences to be considered. For not only is the thread of the story more tangled in the case of Maxwell's theory, but there remain also a number of conceptual loose ends when the thread has been unraveled.

The Maxwell theory clearly reaches its pinnacle with the work of Heinrich Hertz on radio waves. In a series of brilliant experiments Hertz demonstrated the existence of the wide range of electromagnetic waves which had been suggested by Maxwellian theory. The existence of this spectrum of radio waves, in turn, lent strong credence to the view that the spectrum of infrared, visible and ultraviolet light is merely a narrow band in a large spectrum of forms of electromagnetic radiation.

In the course of his researches, however, Hertz hit upon a phenomenon whose exploration was to be a primary stimulus to the development of modern wave mechanics and, accordingly, a primary factor in the overthrow of the classical Maxwellian theory. This was the photoelectric effect. Here is Hertz's description of what he observed:

In a series of experiments on the effects of resonance between very rapid electric oscillations which I have carried out . . . two electric sparks were produced by the same discharge of an induction-coil, and therefore simultaneously. One of these, the spark A, was the discharge-spark of the induction-coil, and served to excite the primary oscillation. The second, the spark B, belonged to the induced or secondary oscillation. The latter was not very luminous; in the experiments its maximum length had to be accurately measured. I occasionally enclosed the spark B in a dark case so as more easily to make the observations; and in so doing I observed that the maximum spark-length became decidedly smaller inside the case than it was before. On removing in succession the various parts of the case, it was seen that the only portion of it which exercised this prejudicial effect was that which screened the spark B from the spark A. The partition on that side exhibited this effect not only when it was in the immediate neighborhood of the spark B, but also when it was interposed at greater distances from B between A and B. A phenomenon so remarkable called for closer investigation.[38]

What piqued Hertz's curiosity here was the appearance of an almost miraculous transmission of electrical energy from one source to another. "For some time, indeed, I was in doubt whether I had not before me an altogether new form of electrical action at a distance." [39]

---

[38] Heinrich Hertz, *Electric Waves*, tr. by D. E. Jones (London: Macmillan, 1900), p. 63. Originally published in *Sitzungsberichte der Berliner Akademie der Wissenschaft*, June 9, 1887. See also, *Wiedemann's Annalen*, Vol. 31, p. 983.

[39] *Ibid.*, p. 4.

This basis for puzzlement disappeared, however, when careful experimentation determined that it was the passage of ultraviolet light from spark A to spark B which produced the anomalous lengthening of the latter. This fact established, Hertz returned to his original line of investigation. But not before he had noted the need for a detailed explanation of the effect in terms of the Maxwell theory. It seems evident from his treatment of the matter that he believed, once the agency of ultraviolet light had been identified, it would be possible to provide a complete account in terms of the Maxwell equations and subsidiary physical assumptions. Certainly he does not indicate any doubt at this point that the theory is correct.

Others, of course, had noted photoelectric effects before Hertz.[40] They had been recorded, however, as mere oddities. The importance of Hertz's findings is that they come at a time when the question of the compatibility with a developed theory of radiation can be raised. And this provides his observations with an historical and logical significance not shared by earlier observations. Others noted stray facts. Hertz recognized the probable presence of a real anomaly.

Philipp Lenard and other investigators began shortly after Hertz's work was made public to try to make good on the explanation Hertz had suggested for the effect (i.e.,

---

[40] Becquerel appears to have been the first. See V. K. Zorykin and E. G. Ramberg, *Photoelectricity* (New York: John Wiley and Sons, 1949), Chapter I; E. Becquerel, "Studies of the Effect of Actinic Radiation of Sunlight by Means of Electric Currents," *Comptes Rendus,* Vol. 9 (1839), pp. 145–49; and E. Becquerel, "Notes on the Electric Effects Produced Under the Influence of Sunlight," *ibid.,* pp. 561–67.

to provide a detailed account of the process by which the ultraviolet light produced the observed effect).[41] Difficulties of a high order followed almost at once. Not only were problems encountered in achieving an adequate theoretical account but there were also a number of misleading and erroneous experimental results to further cloud the picture. We shall return to some of these in a moment.

The well-known work of J. J. Thomson[42] showed in 1899 that the entities emitted in photoelectric interactions of light with metal—including uncharged metal plates— were the very same electrons which made up the cathode ray. This provided firm confirmation of the assumption that the effect was due to interaction of the light waves with the atomic or subatomic constituents of the metal surface. It also provided a more definite framework within which analysis of the photoelectric effect might be carried on.

Following a suggestion put forward by Elster and Geitel in 1890,[43] Lenard proposed[44] that the photoelectric effect is produced when the light wave triggers a process in which an atom becomes dynamically unstable and loses an electron. The trigger action consists of resonance being set up in the atom by a half-wave of the light. On this view, it

---

[41] For instance, Hallwachs, *Annalen der Physik,* Vol. 33 (1888), pp. 301–12; and Elster and Geitel, *Wiedemann's Annalen,* Vol. 41 (1890), p. 175.

[42] J. J. Thomson, in *Philosophical Magazine,* Vol. 48 (1899), p. 547.

[43] Elster and Geitel, *op. cit.*

[44] P. Lenard, *Annalen der Physik,* Series 4, Vol. 2 (1900), p. 359 and Vol. 8 (1902), p. 149.

should be noted, the energy of emission of the electron from the atom is supplied primarily by the atom itself and not by the light wave.

Lenard was led to propose such a mechanism partly because of an extraordinary and anomalous experimental result he had achieved. This result showed that the velocity of emission of the electrons is independent of the intensity of the incident light. Such a conclusion strongly suggested that the mechanism of energy transfer does not involve a straightforward impact of electron and wave front, since the velocity imparted to the electron by such a wave-particle collision would increase with the increase in energy of the wave, i.e., the intensity of the light.

Lenard's attempt to resolve the anomaly *within* Maxwellian theory was somewhat awkward, to be sure. The natural assumption within that theory would be that the energy transmission proceeds directly from the incoming wave to the electron itself. Certainly no other "triggered instability" processes comparable to the one suggested by Lenard were known to exist. Nevertheless, the Lenard proposal—like Hertz's earlier work—offered an "explanation sketch." And the assumption was that the details might be filled in as further empirical evidence became available.

Thomson, meanwhile, was being guided to an entirely different conclusion by another set of experimental data. On the basis of his own work and that of certain predecessors, Thomson came early to the view that increase in temperature intensified the photoelectric action and accordingly led to an increase in the emission velocities of the electrons expelled from the metal surface. With this mistaken view as a premise, he argued in the first edition of his classic work on *Conduction of Electricity through*

Gases[45] that the photoelectrons must be metallic or "free" electrons in the metal—since only these are excited in their motions by heat. He claimed further that their emission was effected directly by action of the wave of light. This account, in contrast to Lenard's, requires that all or nearly all of the energy needed for liberation of the electron from the surface must come from the incident light.

Now in 1907, Millikan and his aide Winchester showed [46] conclusively that Thomson's basic assumption—that temperature affected the photoelectrons' velocities of escape—was based on misleading experimental data: data drawn from experiments conducted in air, rather than in a high vacuum, and data relating to special and unusual cases. In a careful survey of eleven fundamental metals no dependency whatever on temperature was found. This fact, together with Lenard's discovery about the independence of electron velocities and intensity of incident light, led Millikan and his colleague to say: "The independence of the photoelectric effect upon temperature constitutes very conclusive evidence that, if free electrons exist at all within metals, it is not these electrons which escape under the influence of ultra-violet light." [47] Thomson's view, in other words, was untenable. And as a final check, Lenard's result on velocity and light intensity was reconfirmed for all eleven metals. Millikan and Winchester conclude with a

---

[45] J. J. Thomson, *Conduction of Electricity through Gases*, first edition (Cambridge: Cambridge University Press, 1902).

[46] Millikan and Winchester, in *Philosophical Magazine* (6), Vol. 14 (1907), p. 201.

[47] *Ibid.*, p. 198.

remarkable non sequitur. "This result, taken in connexion with the independence both of the discharge-rate and of the velocity of projection upon temperature, seems to establish Lenard's view as to the mechanism of the emission of the electrons." [48]

By the time Millikan and Winchester had arrived at this conclusion, however, two things of great importance had already occurred. For one, Thomson had dropped his earlier hypothesis and had joined Lenard's camp.[49] Millikan's refutation had not been necessary. For another, Einstein's epoch making paper on the "quantum hypoth-

---

[48] *Ibid.*, p. 202.

[49] J. J. Thomson, *Conduction of Electricity through Gases,* second edition (Cambridge: Cambridge University Press, 1906), p. 278: "Lenard has made the very important discovery that the velocity of projection of the corpuscles projected through the agency of ultraviolet light is independent of the intensity of the light. The *number* of corpuscles emitted is proportional to this intensity but the velocity of each corpuscle depends only upon the nature of the illuminated surface. A little consideration shows that this result has very important consequences. It proves that the velocity of the corpuscles is not due to the direct action upon the corpuscle of the electric force, which according to the Electromagnetic Theory of Light occurs in the incident beam of light. It suggests that the action of the incident light is to make some system, whether an atom, a molecule or a group of molecules, in the illuminated system unstable, and that the systems rearrange themselves in configurations in which they have a much smaller amount of potential energy. . . ." It was after this period of tentative acceptance of the Lenard hypothesis that Thomson advanced his aether-string model of the quantized wave front.

esis" had appeared,[50] starting Lenard's account on its way to oblivion. Millikan's argument did not establish what it seemed to establish.

Einstein's 1905 paper swept before it much of the debris which at the time littered the frontiers of research on the nature of light and radiation. Generalizing earlier analyses of the black body radiation problem,[51] Einstein conjectured that all absorption and emission of radiant energy occurs in discrete packets or quanta of magnitude hv,[52]

---

[50] Albert Einstein, "Ueber einen die Erzeugung und Verwandlung des Lichtes betreffenden heuristischen Gesichtspunkt," *Annalen der Physik,* Series 4, Vol. 17 (1905), pp. 132–48.

[51] It is commonly believed that Einstein's work is a generalization of Planck's quantum hypothesis for the black body problem. Martin Klein, however, presents convincing evidence that this belief is mistaken. Klein points out that in his second epoch making paper on light emission and absorption (Einstein, "Theorie der Lichterzeugung und Licht-absorption," *Annalen der Physik,* Series 4, Vol. 20 (1906), pp. 199–206.) Einstein concedes having earlier regarded Planck's hypothesis as radically different from and perhaps inconsistent with his own. There is also the fact that Einstein does not employ Planck's constant in the 1905 paper but scrupulously uses another more complicated expression, apparently to dissociate his hypothesis from Planck's. On this point see footnote 52 below. For Klein's argument see his "Einstein's First Paper on Quanta" in *The Natural Philosopher* (New York: Blaisdell Publishing Co., 1963), Vol. 2, pp. 59–86.

[52] Einstein, as noted above in footnote 51, does not employ Planck's constant in his original paper but instead uses the (equivalent) expression $(R/N)\beta$ where R is the universal gas constant, N is Avogadro's number (the number of molecules

where h is Planck's constant (derived from black body radiation work) and v is the frequency of the radiation absorbed or emitted. When applied to the photoelectric problem, this leads to the conclusion that the energy of an escaping electron ($= \frac{1}{2}mv^2 = Ve$) should equal $hv - p$, where p represents the amount of work done in liberating the electron from the body of the piece of metal.

The relationship postulated by Einstein between frequency of incident light and energy of emission had, to some extent, been noted earlier experimentally. Hertz, for instance, was aware that light of lower frequency than the ultraviolet could be screened out without markedly impairing the spark elongation effect.[53] Not until 1912, however, was it clearly demonstrated that the functional dependence was linear—as Einstein had predicted. And it was 1916 before Millikan's experiments confirmed once and for all that Einstein was right in positing Planck's constant as the constant of proportionality.[54]

The mechanism by which the discrete energy transmissions were presumed to occur was at this point shrouded in obscurity. Einstein himself had failed to produce an acceptable model to accompany his equations. And the

---

in a mole), and β is a constant which appears in Planck's and Wien's distribution law for the spectral density of black body radiation. He thus purports to use Planck's experimental data without using his quantum hypothesis.

[53] Hertz, *op. cit.*, pp. 76–77.

[54] Other important experimental work, meanwhile, had been carried out by Hughes, Richardson, and Compton. See Max Jammer, *The Conceptual Development of Quantum Mechanics* (New York: McGraw-Hill, 1966), p. 35.

Lenard account went down in flames in 1913 when Marx
succeeded in showing that

> . . . an electron, treated as a classical linear oscillator
> in resonancy with the incident radiation, could acquire
> the quantum of energy necessary for its release only if
> the radiation field was 5×10^{11} times as strong as radia-
> tion fields (6×10^{11} watt/meter^2) which had been ob-
> served to yield photoemission.[55]

Two things, however, were clear. First, Einstein's basic
assumption about the "quantized" character of radiant
energy was incompatible with the electromagnetic wave
theory of light. For this theory calls for a uniform distribution of energy in the wave front. Second, Einstein's basic
assumption had been scandalously successful in predicting
all sorts of important facts in the most diverse realms of
radiation physics. In terms of the standard modern treatments of inductive logic, Einstein's hypothesis had obviously piled up the number and variety of "confirmation
instances" to render it highly probable. The Maxwell
theory, except as an heuristic or calculating tool, would
have to be abandoned. It could no longer be construed as
describing the real nature of light. And so Hertz's request
for a Maxwellian explanation of the effect he had observed
leads via a devious route to the downfall of the Maxwell
theory itself.[56]

---

[55] Zorykin and Ramberg, *op. cit.,* p. 25, and Marx, *Annalen der
Physik,* Vol. 41 (1913), pp. 161–90.

[56] In 1938 Einstein and Infeld described the downfall of the
theory of electromagnetic waves in somewhat more straight-

It is obvious that in the sequence of events we have been describing there is a profoundly different order of investigation than that involved in either Newton or Leverrier's work. And there is a sense of vagueness in the outcome which is bound to be disconcerting. Maxwellian electromagnetic theory goes out with a whimper, not a bang.

Look again at the order of the inquiry: Hertz observes the effect, marks it as anomalous but succeeds in provisionally explaining the phenomenon as an effect of light impingement. Other investigators, seeking to fill out Hertz's "explanation sketch" encounter serious problems, though the Lenard account *seems* to carry the day. Then, like a bolt of lightning, Einstein's brilliant conjecture lights up the landscape and provides illumination where previously there had been murkiness. The photoelectric effect is "explained" in such a way as to enable experimenters to detect minutiae which had previously eluded observation. Ein-

---

forward (though historically less accurate) terms: "We should . . . expect the velocity of the emitted electrons to increase if the intensity of the light increases. But experiment . . . contradicts our prediction. Once more we see that the laws of nature are not as we should like them to be. We have come upon one of the experiments which, contradicting our predictions, breaks the theory on which they were based. The actual experimental result is, from the point of view of the wave theory, astonishing. The observed electrons all have the same speed, the same energy, which does not change when the intensity of the light is increased. This experimental result could not be predicted by the wave theory. Here again a new theory arises from the conflict between the old theory and experiment." A. Einstein and L. Infeld, *The Evolution of Physics* (New York: Simon and Schuster, 1938), p. 274.

stein not only offered explanatory understanding, he provided fruitful predictions as well.

*The novel feature here is that at the time of Einstein's
paper there was no demonstrated need for his explanation.*[57] The Lenard theory was still fully credible and

---

[57] Philipp Frank, in his *Philosophy of Science* (Englewood
Cliffs, N. J.: Prentice-Hall, Inc., 1957), pp. 197–99, gives an
entirely different account of the matter. According to Frank,
Lenard's 1902 experiments played the role of "crucial experiment" in destroying the electromagnetic wave theory. Einstein
is then pictured as stepping into the breach to provide an explanation for the already well-defined anomaly. Thus (pp. 198–
199) Lenard's experiment "eliminated" the "undulation theory
in its classical form given by Fresnel, and proved that a 'particle theory' is possible. . . . When Einstein, in 1905, made
the point that this 'second crucial experiment' 'eliminated'
the undulation theory, he attempted to modify it as little as
possible. However, he had to introduce such alterations as
would put the 'renewed' undulation theory in agreement with
Lenard's experiment. From the undulation theory in its classical form, it has been concluded that the energy of vibration
over a spherical wave surface has the same value for equal
areas, but this value decreases as the distance from zero increases. This result is refuted by Lenard's experiment. . . .
The Lenard experiment showed that the energy of each parcel
[quantum in the wave front] is proportional to the frequency of
light: $E = h\nu$, where E is the energy of one parcel, ν the frequency of the light supposed to be monochromatic, and h a
universal constant, called Planck's constant." Several comments
on this incredible passage are in order: (1) Lenard hardly regarded his experiment as "eliminating" the undulatory theory.
On page 170 of his 1902 paper (*op. cit.*) he remarks that although the simplest form of his resonancy hypothesis will not

generally accepted. Even in 1907 Millikan believed that the Lenard hypothesis, in its general outline, was adequate to explain the facts. If this were so, it would have been superfluous to apply Einstein's quantum hypothesis to photoelectricity (though his equation might be accepted as an "empirical law" to be accounted for by Lenard's resonance mechanism). Only gradually as the evidence

---

do, nevertheless there remains to be considered the assumption of more complicated forms of motion within the atom giving rise to states which when triggered by a half wavelength of light lead to the emission of an electron at the appropriate initial velocity. (2) Einstein makes no such point in his 1905 paper as Frank alludes to. He is most cautious about comparing his results with those of Lenard and claims merely a gross agreement. The very title of his paper is intended to suggest the tentativeness of his quantum hypothesis ("heuristischen Gesichtspunkt") and Lenard's hypothesis is not dismissed outright but merely passed over lightly in this manner: "Die übliche Auffassung, dass die Energie des Lichtes kontinuierlich über den durchstrahlten Raum verteilt sei, findet bei dem Versuch, die lichtelektrischen Ersheinungen zu erklären, besonders grosse Schwierigkeiten, welche in einer bahnbrechenden Arbeit von Hrn. Lenard dargelegt sind. Nach der Auffassung, dass das errengende, Licht aus Energiequanten von der Energie $(R/N)\beta v$ bestehe, läast sich die Erzeugung von Kathodenstrahlen durch Licht folgendermassen auffassen." (*op. cit.*, p. 145.) Finally, (3) if Frank were correct in his description of the results of Lenard's experiment *vis-à-vis* the formula $E = h v$, it would be extremely difficult to understand why Millikan subsequently spent ten years of his professional life seeking to check out Einstein's equation for the photoelectric effect. Lenard's 1902 paper, of course, contains nothing remotely resembling the Einstein formula.

seeps in does the need for Einstein's hypothesis in the explanation of photoelectricity become clearer.

It is vital to our understanding of the nature of Einstein's contribution, and our understanding of the logic of explanation as well, to see that without a demonstration of the need for a quantum explanation of photoelectricity Einstein's proposal is logically incomplete. The proposal's acceptability as an explanation hinges on our being sure that *all* alternative hypotheses formulable within classical theory have been ruled out. *Strictly speaking, Einstein's explanation of photoelectricity is not an explanation at all (i.e., is not a satisfactory explanation) until the Maxwellian alternatives have been exhausted.* This fact is at least part of the reason why universal acceptance of the Einstein quantum hypothesis was so slow in coming even after it had been shown to lead to striking unification of many experimental data.

We have here, therefore, a case in which the anomaly is clearly recognized as such only *after* the proposal of the explanation. In other words, we have a case in which the genetic order of development is "hypothesis formulation, prediction, and then confirmation." The full process previously referred to as "defining the anomaly" never seems to be completed. Or at least it does not seem to be carried out prior to general acceptance of Einstein's hypothesis. After that, of course, the question whether the photoelectric effect was *really* anomalous becomes largely academic. And yet the possibility of a clear-cut demonstration of the absolute incompatibility of Maxwellian electromagnetic theory and the photoelectric effect has never been seriously doubted. Once the quantum hypothesis establishes itself in other areas, once its success in explaining photoelectricity is assured, there is no further practical need to belabor the defects in Maxwell's theory. Beating a dead horse is useless effort, and so usually is the activity of defining an anomaly which has already been successfully explained.

In a limited sense, of course, Einstein knew that the photoelectric effect was anomalous. He was aware that there was a "difficulty" in reconciling the effect with the idea of continuous energy distribution in the wave front of light. The problem, briefly put, is that *if* the emitted electrons are "metallic" or free electrons (as Millikan and others were assuming) then emission velocity must depend on intensity of the impinging light—contrary to Lenard's experiment. Einstein's explanation, however, goes far beyond what is required by this "difficulty." Perhaps, after all, Lenard was right in supposing that the electrons were bound within the atom, rather than free. Perhaps they were emitted by a kind of intra-atomic "explosion" rather than by being bumped out by the incident light. How could one be sure? Without hard empirical evidence it was impossible to say. The quantum hypothesis might well be superfluous. This was surely one of the factors which prompted Einstein in 1905 to label his quantum hypothesis an "heuristic standpoint" rather than a new, non-Maxwellian theory of radiation. The photoelectric effect was anomalous, but not *demonstrably* so.

Logically speaking, the vindication of Einstein's boldness comes later—particularly when the Lenard explanation is found to be untenable. This is a crucial point for our understanding of the nature of anomalies. Not only does it show once again the theory-dependency of natural anomalies but it points up also the fact that *the appropriateness and adequacy of an explanation depends on the logical status of the anomaly itself.* The provisionality and "iffiness" of Einstein's explanation of the photoelectric effect arises directly from the lack of a theoretical demonstration of the anomalous character of the effect. It is as though one cannot give a completely satisfactory explanation unless the anomaly is known with certainty to be such. This matter is of vital importance and will be pursued at length in Chapter 2.

For the present, let us conclude our discussion of the Einstein quantum hypothesis with the following historical footnote.

Einstein's own philosophy of science and conception of scientific methodology were surely influenced by the nature of his success with the quantum hypothesis. His repeated insistence that there is no logical route from the data to the hypothesis, that one must propose the hypothesis freely and work out its consequences in order to put it to the empirical test[58]—these doctrines are natural generalizations from the methodology of his work both in the early quantum theory and in relativity.

There is a remarkable parallel in this to Newton's success in the 1672 optical paper and his subsequent adoption of a "hard" empiricist line on methodology. Einstein suc-

---

[58] See, for example, Einstein's "Autobiographical Notes," in *Albert Einstein, Philosopher-Scientist,* ed. by P. A. Schilpp (New York: Harper and Brothers, 1949), Vol. 1, p. 89: "A theory can be tested by experience, but there is no way from experience to the setting up of a theory." And in Einstein's *The World as I See It* (New York: J. J. Little and Ives, Co., 1933), p. 36: "I am convinced that we can discover by means of purely mathematical constructions the concepts and the laws which furnish the key to understanding of natural phenomena."

ceeds scientifically by "free-associating" a spectacular hypothesis and deducing its consequences. Newton succeeds by beginning "from the Phenomena" and working his way to the explanation. And in methodology they conclude by opting for the most disparate viewpoints in the history of scientific thought: *hypotheses non fingo* on the one hand and science as a free creation of the human mind on the other.

Our contention in what follows will be that Newton's and Einstein's descriptions of the methodology of scientific explanation are alike partial and one-sided views of a more fundamental logical model of scientific explanation. Neither is essentially wrong in what he says, but both adopt too narrow a set of limitations on methodology. Either approach can succeed, as their own early works clearly show. And the two methodologies, contrary to the apparent beliefs of Newton and Einstein themselves, are not actually mutually exclusive.

2

Anomalies and the Logic of Scientific Explanation

THE DEDUCTIVE-NOMOLOGICAL PATTERN

Hertz thought the Maxwell theory should be able to explain the photoelectric effect; history records that he was wrong. Einstein, in the end, succeeded in explaining it. Leverrier explained the behavior of the planet Uranus but failed to clear up the key perplexity in the behavior of Mercury. Newton, as Mach put it, "explained facts"— the fact of the elongation of the prismatic spectrum, especially. But what, logically speaking, does it mean to say all of this? What precisely is "explanation"? In Chapter 1 we assumed—quite without justification—that to explain is simply to provide "the cause" of an event. This assumption must be examined more closely. In order to clarify what is meant by 'anomaly' we must also get clear about the meaning of 'explanation.' The two concepts are intimately correlated.

To provide an explanation in the physical sciences is to provide a chain of reasoning in which the anomalous fact or event is linked to (a) prior and/or contemporary

61

events and (b) accepted laws and principles. As Professor
Braithwaite notes, this much is a truism—though its consequences are far from trivial.[1] Among contemporary logicians it has been elaborated into a full-blown picture of
the logic of scientific explanation: what Professor C. G.
Hempel calls the Deductive-Nomological Pattern.[2]

In a classic formulation of the Deductive-Nomological
(D–N) Pattern or Model, Hempel and Paul Oppenheim
lay down four requirements intended to characterize the
nature of sound explanation.[3] While partly aimed at ex-
plicating the idea of "causal explanation" the Hempel-
Oppenheim theses also have a deeper significance. They
seek not only to clarify the meaning of explanation in
contexts where the word 'cause' seems to have some appropriateness but also in cases where it is clearly irrelevant.
Methodologically, at least, it is an oversimplification to
say that scientists explain by "seeking the causes of phe-

---

[1] R. B. Braithwaite, *Scientific Explanation* (New York: Harper
and Brothers, 1953, 1960), p. vii. Braithwaite's formulation of
the truism is slightly different from our own: "It is almost a
platitude to say that every science proceeds, more or less explicitly, by thinking of general hypotheses, of greater or less
generality, from which particular consequences are deduced
which can be tested by observation and experiment. But the
implications of this view are by no means platitudinous."

[2] C. G. Hempel, *Aspects of Scientific Explanation* (New York:
Free Press, 1965), p. 335ff.

[3] Carl G. Hempel and Paul Oppenheim, "The Logic of Explanation," in *Readings in the Philosophy of Science,* ed. by
H. Feigl and M. Brodbeck (New York: Appleton-Century-
Crofts, 1953), pp. 319–52. Reprinted from *Philosophy of Science,* Vol. 15 (1948).

nomena" and on this point the Deductive-Nomological Model sets the record straight. Einstein, after all, was not looking for some empirical condition which led to the photoelectric effect; all of the empirical conditions were known. What he sought was a set of laws and principles of general applicability which would deal with the phenomenon in a perfectly general way. The aim was not to locate experimental conditions but to uncover the true nature of energy transfer processes. Granted: causes and effects are involved. But it is the network of theory— the so-called "nomological" network of laws and principles —with which theoretical scientists are normally preoccupied. Causal talk, in this context, is beside the point.[4]

The main features of the D–N model, as we said, are set out in the form of four requirements for an adequate scientific explanation by Hempel and Oppenheim. These four conditions are supplemented by a remark on the relations between explanation and prediction which has become known as the "symmetry thesis." Following Hempel and Oppenheim we shall adopt the nomenclature '*explanans*' for sentences making up the body of the explanation and '*explanandum*' for the sentence setting out the anomaly.

---

[4] "All philosophers, of every school, imagine that causation is one of the fundamental axioms or postulates of science, yet, oddly enough, in advanced sciences such as gravitational astronomy, the word 'cause' never occurs." Bertrand Russell, "On the Notion of Cause . . ." in *Readings in the Philosophy of Science, op. cit.,* p. 387. (Reprinted from *Our Knowledge of the External World* (New York: W. W. Norton and Co., 1929), p. 247.) Which is not to say, of course, that gravitational astronomy is not a causal theory.

Of the requirements for adequate scientific explanation proposed by Hempel and Oppenheim three are listed as "Logical Conditions of Adequacy" (R_1–R_3) and the fourth is termed an "Empirical Condition of Adequacy" (R_4). They are as follows:

(R_1) The explanandum must be a logical consequence of the explanans; in other words, the explanandum must be logically deducible from the information contained in the explanans, for otherwise the explanans would not constitute adequate grounds for the explanandum.

(R_2) The explanans must contain general laws, and these must actually be required for the derivation of the explanandum. We shall not make it a necessary condition for a sound explanation, however, that the explanans must contain at least one statement which is not a law; for, to mention just one reason, we would surely want to consider as an explanation the derivation of the general regularities governing the motion of double stars from the laws of celestial mechanics, even though all the statements in the explanans are general laws.

(R_3) The explanans must have empirical content; i.e., it must be capable, at least in principle, of test by experiment or observation. This condition is implicit in (R_1); for since the explanandum is assumed to describe some empirical phenomenon, it follows from (R_1) that the explanans entails at least one consequence of empirical character, and this fact confers upon it testability and empirical content. . . .

(R_4) The sentences constituting the explanans must be true.

It is assumed that the "other" statements which constitute the explanans will consist of so-called antecedent condition statements (in the case of straightforward causal explanation) or statements of initial and boundary conditions in the case of theories of mathematical physics. Hempel and Oppenheim throughout their paper speak only of antecedent conditions but in his more recent writings Professor Hempel has generalized the account to include other types.[6] Subsequently we shall have occasion to discuss the nature of boundary and initial conditions at greater length.

Hempel and Oppenheim's four conditions have not, of course, gone unchallenged.[7] And while it is not our purpose here to attempt to evaluate these criticisms (since, as

---

[5] Hempel and Oppenheim, *op. cit.*, pp. 321–22.

[6] Hempel, *Aspects of Scientific Explanation, op. cit.*, pp. 331–489.

[7] See, for example, Israel Scheffler, "Explanation, Prediction and Abstraction," in *Philosophy of Science,* ed. A. Danto and S. Morgenbesser (Cleveland and New York: World Publishing Co. (Meridian Books), 1960), pp. 274–87. Scheffler's article in a more extended form appeared originally in the *British Journal for the Philosophy of Science,* Vol. 7 (1957). See also Michael Scriven's "Explanations, Predictions and Laws," in *Minnesota Studies in the Philosophy of Science,* Vol. 3, ed. H. Feigl and G. Maxwell (Minneapolis: University of Minnesota Press, 1962), pp. 170–230. Professor Scriven has been the most persistent and thorough critic of the Hempel position.

will appear, our concern is not with any one of the four conditions or the symmetry thesis individually), it is nevertheless worthwhile to set out some of the main points of contention.

(A) Condition (R_1) has most often been attacked as an unrealistic limitation on the character of good explanations. According to this view, scientific explanations can very well fail to be deductively valid arguments without thereby forfeiting their scientific credibility or adequacy. Merely pointing to *a* condition which causes the observed effect suffices in many cases. This does not constitute giving a deductive argument.

The typical Hempelian answer to this objection is that such "explanation sketches" are adequate only insofar as they can be filled out to match the deductive requirement. When they cannot be so elaborated from contextual clues they are demonstrably inadequate as scientific explanations. Thus, the deductive requirement stands as a criterion which must at least be satisfiable if an explanation is to be adequate.

Besides, argues Professor Hempel,[8] how can one cite a *single* antecedent condition without implicitly or explicitly calling upon some general law to connect it with the explanandum-event? Causal connections do not exist apart from the nomological or theoretical context of beliefs we bring to the subject matter. There is no "cosmic glue" holding events together. When we assert, therefore, that event A causes event B we must be implying that some general relationship—a law—links the events for us. It follows that a deductive linkage between explanans and

---

[8] Hempel, *Aspects of Scientific Explanation, op. cit.*, p. 350.

explanandum will be found if the explanation is actually a good one.

More recently, Hempel, Grünbaum and others have pointed out that objections to (R_1) are often based on the mistaken assumption that it requires *all* explanations to be deductive in form.[9] As Hempel is quick to concede, there is indeed at least one type of explanation which does not employ deductive inference: namely, inductive-statistical explanation. In this form of explanation it is unnecessary to show that the explanandum-event was predictable with certainty but only that it was to be expected with some degree of probability. The requirement of deductive connection clearly does not apply to these explanations but only to so-called "nomological" or "causal" explanations.

(B) The requirement that at least one general law statement be included in the explanans is likewise challenged on grounds that certain actual explanations do not include such laws. When, for example, we explain Johnny's having contracted measles by pointing out that he had just been in contact with cousin Mathilda shortly before and that Mathilda was at the time coming down with measles, no *general* law concerning measles is invoked. For it is not always (or even for the most part) true that contact with a person coming down with the measles leads to contraction of the disease.

The usual Hempelian answer to this type of objection is

---

[9] C. G. Hempel, *Aspects of Scientific Explanation, op. cit.* and also Adolf Grünbaum, "Temporally-Asymmetric Principles, Parity Between Explanation and Prediction, and Mechanism versus Teleology," *Philosophy of Science,* Vol. 29 (1962), pp. 146–70.

that a confusion of two types of explanation—inductive-statistical and deductive—lies at the heart of the matter. The explanation of Johnny's getting measles is not really a deductive explanation at all but a statistical and inductive one. It rests on the assumed premise that in such and such percentage of cases of exposure to measles the disease is contracted. Thus, the failure of the explanation to rest upon a general law statement *is* evidence that the explanation is not a sound deductive explanation at all.

(C) The persistent difficulties experienced by traditional positivists in specifying the meaning of 'verifiable,' 'testable' and 'empirical content' form the basis for most criticism of Hempel and Oppenheim's (R_3). According to this requirement the explanans statements must be testable via experiment or observation. What precisely does this mean? What are the logical marks of an "empirically meaningful" statement? Here the argument tails off into the familiar discussions of verifiability criteria and the like.[10]

Whether or not such discussions can be terminated by agreement is an open question but not one which seriously affects Hempel's theses on explanation. The general intent of the criterion is clear and except for certain controversial cases no one has much difficulty in understanding what is being required.

(D) Finally, as regards the requirement of truth, Hempel and Oppenheim themselves point out the difficulties.

That in a sound explanation, the statements constituting the explanans have to satisfy some condition of

---

[10] The *locus classicus* of these discussions is Professor A. J. Ayer's justly famous *Language, Truth and Logic* (New York: Dover Publications Inc., 1935, 1946).

factual correctness is obvious. But it might seem more appropriate to stipulate that the explanans has to be highly confirmed by all the relevant evidence available rather than that it should be true. This stipulation, however, leads to awkward consequences. Suppose that a certain phenomenon was explained at an earlier stage of science, by means of an explanans which was well supported by the evidence then at hand, but which had been highly disconfirmed by more recent empirical findings. In such a case, we would have to say that originally the explanatory account was a correct explanation, but that it ceased to be one later, when unfavorable evidence was discovered. This does not appear to accord with sound common usage, which directs us to say that on the basis of the limited initial evidence, the truth of the explanans, and thus the soundness of the explanation, had been quite probable, but that the ampler evidence now available made it highly probable that the explanans was not true, and hence that the account in question was not—and never had been—a correct explanation.[11]

A less rigorous, but perhaps equally satisfactory, way to deal with the matter is to assume that the Hempel criteria deal with the concept of *"the* explanation" rather than *"an* explanation." Thus, in order to have attained *The Scientific Explanation* of a phenomenon one must have arrived at the true laws of nature and the true and complete picture of the antecedent or initial conditions; whereas to have *A Scientific Explanation* is to have an account satis-

---

[11] Hempel and Oppenheim, *op. cit.,* p. 322.

fying R_1–R_3 (and the symmetry thesis) but not necessarily R_4. At any stage in the history of science, we can properly describe mankind as having *A Scientific Explanation* of a particular phenomenon, but rarely—if ever—can it be claimed that *The Scientific Explanation* has been arrived at. Hempel himself distinguishes between "potential explanations" and "true explanations" in somewhat this fashion.[12] To have a potential explanation it is sufficient that R_1–R_3 be satisfied (with the proviso that in R_2 we speak of "law-like generalizations" rather than true laws). Most of the explanations offered in the history of science would be of this type.

Having considered Hempel and Oppenheim's R_1–R_4 in some detail let us now sum up by describing the general features of a sound scientific explanation as compactly as possible: such an explanation consists of a set of statements (explanans) in which empirically meaningful and true statements of two types—general laws and antecedent (initial and boundary) conditions—are included. The set is deductively linked with another statement (explanandum) which describes the anomaly to be explained and which does not follow validly from the explanans without inclusion of all those statements which are present in it.

To these criteria and the concomitant model, Hempel and Oppenheim append a further proviso which has come to be even more controversial than the original four requirements.[13] The proviso is that an explanans is not fully

---

[12] Hempel, *Aspects of Scientific Explanation, op. cit.,* p. 249 (footnote).

[13] For a discussion see N. R. Hanson, "On the Symmetry Be-

adequate to explain the explanandum-event unless the explanans would have sufficed prior to the event as a basis for predicting its occurrence. In other words, any explanation worth its salt must also have predictive power. This is the so-called symmetry thesis.

Much ink has been spilt in criticism of this thesis—some of it on the mistaken assumption that Hempel and Oppenheim were requiring that all scientific predictions count as explanations. The title 'symmetry thesis' is extremely misleading in this respect and is probably responsible for much of the confusion. But all that Hempel and Oppenheim actually claim is that an explanation (if it is adequate) must suffice for predictive purposes. In general, the counterexamples brought against this claim have not been particularly effective though, again, it is an open question among philosophers of science at the moment whether the Hempel-Oppenheim requirement is a legitimate one for areas of science outside classical mechanics and analogous physical theories.

The four basic D–N model criteria plus the symmetry thesis are, in any case, amply illustrated in the three historical examples we dealt with in Chapter 1. Let us consider them in turn:

(1) Newton's explanation of the spreading of the spectrum is a clear-cut case of deductive explanation, Stephen

---

tween Explanation and Prediction," *The Philosophical Review,* Vol. 68 (1959), p. 349, and the rebuttal in Adolf Grünbaum's "Temporally-Asymmetric Principles, Parity Between Explanation and Prediction, and Mechanism versus Teleology," *op. cit.*

Toulmin's odd views on geometrical optics notwithstanding.[14] The laws in this case include the principle of rectilinear propagation of light, and Snell's law in its post-Newtonian form (i.e., with Newton's amendment concerning the differences in degree of refraction of component rays built into the table of indices of refraction). The initial and boundary conditions in this case are expressible in statements describing the apparatus, the type of incident light, and the (observable) path of the beams of light. From this data one literally can deduce the expected distortion in the shape of the image. Indeed, given all of this data beforehand one can actually predict the shape and dimensions of the image before it is produced. Finally, we add that to the best of man's knowledge in Newton's time the laws invoked and the descriptions of conditions offered were actually true. Slight modifications—amendments of the type Newton himself had given—would have to be added in order for us to hold the laws to be true to the best of *present* knowledge. Surely, there is no question that all of Newton's premises are empirically meaningful.

(2) Since Newtonian celestial mechanics is currently believed to involve significant theoretical error, we must leave out Hempel and Oppenheim's (R_4) in discussing Leverrier's explanation of the behavior of Uranus. The true explanation of Uranus' perturbed orbit evidently re-

---

[14] Stephen Toulmin, *The Philosophy of Science* (New York: Harper and Brothers, 1953, 1960), see especially p. 41. For a refutation see May Brodbeck, "Explanation, Prediction and 'Imperfect' Knowledge," in *Minnesota Studies* . . . , vol. 3, *op. cit.*, pp. 231–72.

quires relativistic treatment. Considered in the context of the mid-19th century, however, it is clear that the explanation offered by Leverrier is a paradigm of D–N explanation.

The main factor, historically, in Leverrier's explanans is his description of the initial and boundary conditions: particularly his bold (but somewhat inaccurate) conjectures on the elements of Neptune's orbit and mass. From these data, suitably corrected in light of later observations of Neptune, and using Newton's laws of motion and universal gravitation, one can literally deduce descriptions of the observed behavior of Uranus. The deductive inference is, once again, a mathematical form of inference. And the predictive capacity of the explanans is likewise self-evident. Had astronomers known of Neptune's existence and orbit before Leverrier's day there would have been no serious problems in reconciling theory with observation.

(3) Einstein's explanation of the photoelectric effect even more brilliantly exhibits the Deductive-Nomological pattern, if that is possible. The laws involved include the Einstein quantum hypothesis, from which the Einstein photoelectric equation and hence a description of the effect can be deduced, given a suitable account of the apparatus and other physical conditions. Other principles and definitions are invoked, of course, but these present no special difficulty for the D–N model.

As before, the Einstein explanation has remarkable predictive power. Historically this can be seen by noting how experimenters were led to adopt more refined techniques when predictions from the Einstein equation seemed not to correspond to the facts. When techniques were refined, it was discovered that the exact relation between light frequency and electron velocity predicted by Einstein was in fact the one which obtains in nature.[15] Einstein's predictions using the explanans under consideration were more reliable than the actual early observations!

It can be seen that in each of our three examples the Deductive-Nomological Model provides a suitable logical account of the structure of good scientific explanation. Whether or not the model applies equally well in other areas and branches of science, it does work here. Even so, there is something fundamentally wrong with the picture it provides us of scientific explanation. Something vital has been left out.

ANOMALIES AND EXPLANATIONS

The law for the simple pendulum $[T = 2\pi\sqrt{L/g}]$ makes it possible not only to infer the period $[T]$ of a pendulum from its length $[L]$, but also conversely to infer its length from its period; in either case, the inference is of the form (D–N). Yet a sentence stating the length of a given pendulum in conjunction with the law, will be much more readily regarded as explaining the pendulum's period than a sentence stating the pe-

---

[15] Martin J. Klein, *op. cit.*, pp. 78–79: "The prediction that Einstein made in the photoelectric equation was a bold one, almost as bold as the theory that led to it. Nothing at all was known about the frequency dependence of the stopping potential in 1905, not even the existence of such a dependence, and Einstein was predicting both its form and the precise value of the essential constant in the equation. It actually took almost a decade of difficult experimentation before all features of Einstein's equation could be fully tested."

riod, in conjunction with the law, would be considered as explaining the pendulum's length.[16]

The problem here is that mere subsumption of an event under a law as required by the D–N model of scientific explanation is not in and of itself sufficient to count as explanation. *Something is missing in the D–N model and it is something essential.*

In his recent book, *Aspects of Scientific Explanation,* Professor Hempel ponders this problem and comes to the following conclusion:

> In cases such as this, the common-sense conception of explanation appears to provide no clear grounds on which to decide whether a given argument that deductively subsumes an occurrence under laws is to qualify as an explanation.[17]

But is it really a failure of our common-sense conception of explanation at all? Clearly, the root of the trouble lies in the D–N model itself, not in our ordinary intuitions of what counts as explanation. We can see that the argument based on the pendulum law terminating in the explanandum "The length of the pendulum is L" is simply not an explanation of the fact.[18] Nothing is wrong with our in-

---

[16] Hempel, *Aspects of Scientific Explanation, op. cit.,* pp. 352–3.

[17] *Ibid.,* p. 353.

[18] Later, however, we shall point out how this argument *could* be construed as an explanation in certain very unusual contexts.

tuitions. What *is* wrong is that the D–N model in no way rules out the proffered explanans as inadequate. It passes all of the Hempel-Oppenheim tests.

Many similar examples from the sciences can be produced. Let us for simplicity's sake, however, give a more elementary one in hopes of seeing just what the matter is. For instance,

> All men have noses on their faces.
> I am a man.
> _____
> Therefore, I have a nose on my face.

Surely, one might argue, this does not count as a valid scientific explanation even if it is assumed to meet the Hempel-Oppenheim requirements. It is just not the *kind* of argument we would be inclined to accept as an explanation of the fact. An anti-Hempelian might put it this way: "Anyone who is aware of the fact that he himself has a nose has already become aware of the general nasality of mankind and of his own humanity. He will hardly find his understanding enhanced by this bit of idle syllogizing. The mere presence of a generalization in the explanans does not convey genuine understanding."

Defenders of the D–N Pattern may point out in rejoinder that the alleged "general law" in this case is not true without qualification. Mutations and accidents produce some of the exceptions. Accordingly (the argument continues) if the law is properly stated in terms of the conditions under which men do or do not come to have noses the syllogism here presented will develop into a bona fide scientific explanation of the presence of my nose upon my face.

Both sides in this controversy, I wish to claim, would be

shooting wide of the mark. For one thing, even if it were true that all men have noses, the syllogism would hardly count as a sound scientific explanation. Or, at least, it would not count as such without our first having made some very strange assumptions. On the other hand, the inadequacy of the alleged explanation is not merely dependent on the association of knowledge of the conclusion with knowledge of the premises. *Even if a person were not aware of the information contained in the premises he might still fail to be intellectually satisfied by having them supplied.* For suppose he has in mind the question: "Why do I have a nose on my face *and not a horn like the rhinoceros?*" The answer: "Because you are a man and all men have noses on their faces" will not really solve his problem. Nor, by hypothesis, is this just a case of his asking why *all* men have noses (the usual deductivist gambit at this point). For we have assumed that he does not know at the outset that all men have noses. How then can he be asking for an explanation of that law if he does not know it?

What is evidently lacking in the D–N account is the idea that the explanandum-event or fact must have some special status if it is to function as the object of a scientific explanation. To present the Deductive-Nomological Model as a sufficient account of sound explanation is to resurrect, in a more subtle form, the fallacious image of the string of identical pearls. And this image, as our historical examples have shown, is grossly inaccurate.

Proponents of the D–N account have seldom if ever argued that it presents a set of jointly sufficient conditions for adequate explanation. Professor Hempel, in particular, scrupulously avoids taking this position. The D–N account is for him nothing more than a description of what sound explanations should look like after they have been properly formulated. But this kind of neutrality, I wish to argue, will just not do. The logical and epistemic adequacy of a proposed explanation depends essentially on the anomalous character of the events or facts described by the statement. A model of explanation which purports to judge the adequacy or inadequacy of various explanatory proposals cannot overlook this fact. *If the Hempel D–N model is to be taken merely as a descriptive account of some necessary conditions of sound explanation it can never be used to certify that a proposed explanation is indeed logically and epistemologically adequate.* Thus, in each of the three historical illustrations discussed above we should be obliged to say that although all of the D–N criteria are satisfied we can conclude nothing about the logical and epistemic adequacy of the explanations given. That they meet the specifications R_1–R_4 and the symmetry criterion does not show that they are adequate explanations. Surely this represents an unsatisfactory state of affairs for proponents of the D–N model.

There is, of course, more than one thing missing in the D–N account. As a minimal addition, however, this kind of requirement appears to be essential:

> (R_5)—The explanandum statement must be the description of a naturally anomalous state of affairs or event.

In other words, what is needed is at least a requirement limiting the range of explicable phenomena to those which actually need to be explained.

Certainly this fifth requirement was satisfied in the examples we have considered: the spread of the spectrum,

Uranus' positional aberrations, the lengthening of Hertz's spark, and the emission energy-light intensity relation discovered by Lenard were all natural anomalies. They demanded explanation. The presence of a nose on one's face, on the other hand, is not a fact which normally needs explanation. It would be surprising (to say the least) were one's nose to disappear some morning, but there is nothing really puzzling about its just being there.

As soon as this has been said, however, it becomes evident that the concept of natural anomaly is an extraordinarily fuzzy one. What seemed perfectly clear in light of the historical examples seems to disintegrate on closer examination. Aren't there, after all, puzzling aspects about the presence of a nose on one's face? Isn't it really every bit as much in need of explanation as the photoelectric effect? Almost at once the gestalt pattern of peculiarity replaces the gestalt of naturalness. The line between what requires explanation and what does not becomes perilously blurred.

Before seeking to clarify the idea of anomaly, however, a further point needs to be made about the D–N model of explanation. For the fact is that even with the addition of the fifth requirement to those laid down by Hempel and Oppenheim we still have not got a fully adequate account of explanation in the sciences.

The trouble is that explanation depends not only on the invocation of "covering laws" but also upon the context of beliefs which gives rise to the request for explanation in the first place. This context—we shall refer to it hereafter as "the anomaly context"—determines to a large extent what shall count as a satisfactory explanation. *It is not possible to give an adequate account of scientific explanation without taking context into consideration.*

Consider again Professor Hempel's example regarding the explanation of the length L of a pendulum by reference to its period T. Surely anyone who asks "Why does the pendulum have length L?" will not normally be puzzled about this fact because of beliefs he holds about the pendulum's period. The anomaly context in which his question comes up probably includes no beliefs one way or the other about periods of pendulums. For this reason, the law $T = 2\pi\sqrt{L/g}$ is irrelevant to the resolution of his perplexity. Both psychologically and logically the explanans is inadequate to explain the fact at hand.

It is conceivable, of course, that someone might be led to ask why the pendulum has length L on the basis of beliefs about its period. One might, for instance, believe (incorrectly) that the pendulum has period T', different from T, and be led to expect a length L', different from L. In this kind of context—a very unusual one, to be sure—the deductive argument from the law and initial conditions to length L *would* count as an explanation! Understanding is conveyed by the correction of our mistaken belief about the period. It is virtually a certainty that Professor Hempel did not have in mind this kind of situation when he originally proposed the example.

The kind of condition which is evidently needed in order to take into consideration the role of intellectual context is something like the following:

(R_6)—An explanans must include as many of the assumptions and beliefs of the anomaly context as is consistent with requirements R_1 through R_5. In other words, there must be minimal deviation from the context of beliefs in which the original request for an explanation is cast.

It is apparent that this requirement rules out the pendulum-length explanation for most contexts. Unless one's expectations about the length of the pendulum are based on information about its period the period-length law is probably going to be irrelevant.

According to R_6, an explanans must represent the solution of a kind of conceptual problem posed by the context of beliefs. The significance of this is that it ties explanations firmly to contextual bases. "Explanatory power" and enhancement of understanding are seen not merely as by-products of deductive inference from law statements to descriptions of facts but as the results of a conceptual shift from anomaly context to explanans. To be explained is to be seen in a new light—the light of understanding.

THE RELATIVITY OF ANOMALIES

If our suggestion about the need for additional requirements (beyond those usually given in the D–N model) is to have validity, a more lucid account of the idea of natural anomaly must be given. The key to such an account is already at hand. It is the apparent fact that what shall count as an anomaly is a function of the background of beliefs and knowledge against which the anomaly is set. Anomalies are what they are in virtue of the conceptual setting in which they occur. Here is a formal statement of what all of this means:

Suppose that a self-consistent body of beliefs, facts and hypotheses is given and suppose that it is expressed as a conjunction B of statements. Then we may say that a state of affairs described by the sentence 'A' is *anoma-*

lous with respect to the beliefs expressed by 'B' if and only if the conjunction $(A \& B)$ is self-contradictory.

Thus, the test for anomalousness—in this relative sense—is a purely logical and formal one dependent only upon the logical relation between beliefs held and the state of affairs construed as anomalous.

Nothing essential, it must be emphasized, hinges on the fact that B expresses a set of beliefs. We are concerned here not with the psychology of persons giving or receiving explanations but rather with the logical relations among propositions they are prepared to assert.

Moreover, the characterization of anomalies offered here is strictly a non-pragmatic one. What counts as an anomaly under our definition in no way depends upon the intentions, preferences or purposes a scientist has but only on the beliefs he is prepared to subscribe to. Hence to say that some fact or state of affairs is (in the present sense) anomalous is to make a factual and logical claim, not to express a value judgment about what scientists "prefer" to explain.

It is natural to ask at this point whether the *relative* definition can be absolutized; i.e., whether the notion of an anomaly independent of any conceptual context whatever can be made intelligible. We have as a matter of fact already implicitly answered this question in the negative. *The notion of an absolute anomaly is the notion of an event or state of affairs whose occurrence would be logically impossible to explain.* In point of fact, such an event or state of affairs would not even be describable without self-contradiction. For suppose that the sentence 'p' describes such an event. Then, if we extend the definition given above in a natural fashion to the idea of absolute anomaly, we shall be obliged to say that '*p*' is logically incompatible with any set of statements whatever. This implies that the negation of '*p*' is logically entailed by any set of statements and hence is a tautology. The statement '*p*,' in turn, must be self-contradictory.

Charles Sanders Peirce, one of the few logicians ever to take up the question of what constitute grounds for requesting an explanation in science, held a position somewhat different from the one taken here and it may be helpful for purposes of clarification to consider his views carefully for a moment. Interestingly enough, Peirce—the father of modern pragmatism—adopts a non-pragmatic account of the nature of anomalies. Nothing is made to depend upon the purposes, desires or intentions of persons. According to Peirce,

> . . . the only case [which] . . . leads to the conclusion that an explanation is positively called for, is the case in which a phenomenon presents itself which, without some special explanation, there would be reason to expect would *not* present itself; and the logical demand for an explanation is the greater, the stronger the reason for expecting it not to occur was.[19]

This is fairly close to the notion set out above that an anomaly is an event whose occurrence contradicts a set of beliefs we are entertaining. Peirce, however, comes to the remarkable conclusion that "All conceivable facts are di-

---

19 C. S. Peirce, *Collected Papers,* edited by Charles Hartshorne, Paul Weiss and Arthur Burks (Cambridge: Harvard University Press, 1931–, 8 vols.), Vol. 7, p. 114.

visible into those which, upon examination, would be found to call for explanation and those which would not." [20] In other words, there are anomalies which are absolute and still other facts which absolutely require no explanation.

In some sense, what Peirce is saying is entirely correct. Some events or facts *naturally* demand explanation in a way others do not. But Peirce's view tends to be misleading in failing to allow for the possibility that one can reasonably ask for an explanation of any fact or phenomenon provided only that one has a set of beliefs which make it anomalous. Relative to a given set of beliefs all facts are divisible into those which require explanation and those which do not. But since there is no one unique set of beliefs shared by all men it must be allowed that any fact or phenomenon can be *seen as* requiring explanation. Peirce explicitly denies this. "[Some philosophers] probably would say, that all facts call for explanation more or less. According to me, however, the demand for explanation is a more definite demand." [21] Apparently, Peirce is seeking to formulate a non-relativistic concept of anomaly based on the idea that the reasons supporting the request for an explanation are either good or bad ones (hence: the request itself is either legitimate or illegitimate). We shall explore this more fully in a moment.

Reverting once more to our historical examples, the applicability of the definition of 'anomalous with respect to' given above is easy to see. In every case, a well-defined

---

[20] *Ibid.*, Vol. 7, p. 117.
[21] *Ibid.*

body of physical theory backs up the identification of a particular fact as anomalous. Newton's spectrum, for example, is oblong when, according to Descartes' (Snell's) law, it could be expected to be round. By means of careful experimentation and reasoning Newton succeeded in showing that the elongation was logically impossible given the existing theory of geometrical optics. Leverrier's painstaking recalculations of the Bouvard tables were likewise devoted to showing the logical incompatibility of observation and theory. In this case, the theoretical background is that of celestial mechanics plus the catalogue of earlier observations on the known bodies of the solar system.

The photoelectric effect actually exhibits a doubly anomalous aspect when considered from the present point of view. As Hertz saw it, the effect was anomalous since it appeared to contradict contemporary beliefs about the number of types and the forms of action at a distance. When the phenomenon was traced to the electromagnetic radiation of ultraviolet light, Hertz was at least partially content. He had found what Hempel speaks of as an "explanation sketch" and the presumption was that this sketch could be elaborated into a deductive explanation of the Hempelian type.

In contrast to the slightly superficial anomalousness which originally attracted Hertz's attention, the photoelectric effect was later found to involve a deeper incompatibility with the theory of electromagnetic waves. As we have seen, this incompatibility did not become apparent until after Einstein's explanation had been put forward. In this instance the anomalousness stemmed from two facets of the observations of the effect. A present day physicist describes the situation as follows:

Is the electromagnetic theory of light capable of explaining the photoelectric effect? At the first glance it seems possible. In this theory electric and magnetic fields are attributed to the light waves, which may well exert forces on the electrons contained in the metal and so liberate them from the metal surface. However, on this basis we should predict that light of high intensity, i.e., consisting of strong fields, would give high kinetic energy to the electrons liberated. This is not the case; on the contrary, experiments show that the kinetic energy is independent of the light intensity. Furthermore, according to the theory we should expect that light of low frequency (infrared), if sufficiently intense, would be as effective as high-frequency light (ultraviolet). Again this prediction of the electromagnetic theory contradicts the experiments, which show that light of a frequency below a sharply defined threshold is entirely ineffective. So we are forced to conclude that the electromagnetic theory of light fails to explain the photoelectric effect.[22]

The logical pattern here is essentially as before. Electromagnetic wave theory plus initial and boundary conditions imply two observable relationships which do not in fact occur in nature. Hence, descriptions of the observations logically contradict the body of accepted theory and experimental data.

It should be somewhat easier to see at this point what was involved in the process we spoke of in Chapter 1 as

---

[22] Otto Oldenberg, *Introduction to Atomic and Nuclear Physics,* 3rd edition (New York: McGraw-Hill, 1961), p. 66.

"defining the anomaly." The process amounts logically to a formal demonstration of the incompatibility of theory and observation. This typically involves a considerable degree of empirical research on relevant conditions affecting the putative anomaly. For in order to establish once and for all that theory and observation are incompatible it must be determined with near certainty that theory plus descriptions of the relevant conditions definitely imply the non-occurrence of the event which actually occurs.[23] It is usually a matter of the utmost experimental and observational delicacy. And in view of the fact that the work done in defining an anomaly (e.g., that of Newton, Leverrier and Millikan) is normally cited by all later theoreticians as the "last word" on the subject, the importance of the procedure of defining anomalies for the growth of scientific knowledge cannot be overestimated.

Having analyzed the concept of "anomalous with respect to" let us now turn to an attempt at clarifying the more important notion of "natural anomaly." We shall attack the problem broadly, i.e., from the standpoint both of scientific and non-scientific explanation, restricting our attention to scientific explanation thereafter.

We have already conceded that any actual event, however ordinary, can be construed as an anomaly if one is prepared to undertake the necessary intellectual contor-

---

[23] An index of the importance of this activity to actual scientific practice is given by George Gamow: "Staggering contradictions . . . between theoretical expectations on the one side and observational facts or even common sense on the other are the main factors in the development of science." Gamow, *Thirty Years that Shook Physics: The Story of Quantum Theory* (Garden City, N. Y.: Doubleday, 1966), p. 37.

tions. On the other hand, we have tried to depict a *natural* anomaly as one whose anomalousness stares us in the face. What precisely is the difference here? What do we mean when we speak of "intellectual contortions"?

What, for instance, is it necessary to do in order to view the presence of a nose on one's face as an anomaly? One thing that is necessary is the adoption of a set of beliefs or opinions with respect to which the occurrence is anomalous. That is to say, a set of beliefs which imply that one's nose should either not exist or not be present on the face.

It is not difficult to imagine such sets of beliefs. For example, if a guillotine or knife had just taken off my eyebrows, part of my lips and the tip of my chin in a single, swift pass, I should be very surprised to have a nose left on my face. Or if I had just seen an elephant or anteater for the first time and were inclined to think of myself as belonging to essentially the same species as one of them, my nose would be a bit bewildering.

Any number of examples could be given. The crux of the matter, however, is this: *every set of beliefs one can propose involves at least one belief which does not begin to have the presumption of truth on its side.* The assumptions are either deliberately counterfactual and hypothetical or else are explicitly false and presumed true on the basis of incredible ignorance. What was previously referred to as "intellectual contortions" is just the process of making such counterfactual assumptions or assumptions of extreme ignorance.

What, then, is a naturally anomalous state of affairs? It is one which we have at least plausible reasons to believe should not have come to pass. In other words, it is a *state of affairs anomalous with respect to a set of statements which are at present putatively true.* In the case of scien-

tific anomalies this means: anomalous with respect to existing theory, observation reports or reports of experimental
data. Newton's elongated spectrum, Uranus' strange movements, the odd relation between frequency and escape
velocity in the photoelectric effect—all are naturally anomalous in this sense.

There is another more direct way of putting the matter:
*any fact or state of affairs which actually requires explana-
tion can be shown to be in need of explanation on the basis
of existing knowledge.* A plausible argument, founded
upon already accepted principles and facts, can always be
constructed in order to show that the fact at hand is incompatible with those accepted principles and facts. And
it is even logically legitimate to demand such a proof when
an explanation is requested. "Why shouldn't it have hap-
pened?" "What's so odd about that?" "What else did you
expect?" These are questions designed to elicit knowledge
of the background against which the event or fact is being
viewed. And they are at the same time requests for justification of the alleged need for an explanation. When the
person requesting an explanation cannot answer such questions as these it is very unlikely that we can produce a satisfactory explanation for him.

In scientific contexts, at least, it is always reasonable to
ask for justification of an explanation request. This, as we
shall see, is a direct consequence of the fact that in scientific discourse the concept of "present state of knowledge"
has a clear and unambiguous meaning. Even outside of
science, however, it is often both useful and clarifying to
inquire into the presuppositions of why-questions intended to elicit an explanatory answer. Occasionally one
finds that the question depends for its plausibility on self-
contradictory assumptions. For instance, when one asks—

as metaphysicians often do—why there should be something rather than nothing at all, one presupposes a set of beliefs implying the non-existence of the universe and all it contains. A little reflection shows that any such belief set, insofar as it gives *reasons* for expecting the world to be non-existent, will imply the existence of something or other and hence be inconsistent.

It is a remarkably embarrassing fact about the D–N model of explanation that it allows statements like 'Something exists' to stand on a par with descriptions of the photoelectric effect as matters worthy of explanation.

The idea that the request for an explanation can always be justified is apparently the basis for C. S. Peirce's theory of anomalies discussed earlier. Peirce, it will be recalled, held that events may be divided neatly into two classes: those requiring explanation and those not requiring it. The division seems to depend upon whether the request for explanation is justified by sufficient reasons or not. ". . . the logical demand for an explanation is the greater, the stronger the reason for expecting it [the anomaly] not to occur was." [24] It is difficult to see, however, exactly where Peirce wishes to draw the line between sufficient and insufficient reasons.

The idea of natural anomaly—as set out above—draws the line fairly sharply. Or, at least, as sharply as seems warranted by ordinary usage. Yet it does not require that all events be divisible into two classes for all time. *A natural anomaly is anomalous with respect to existing knowledge at a particular time.* It is still, in this sense, a surprising phenomenon relative to some body of beliefs, not an

---

[24] Peirce, *op. cit.*, Vol. 7, p. 114.

"absolute surprise." Peirce's "events requiring explanation," on the other hand, seem to be independent of particular sets of beliefs or the state of human knowledge at a
particular time. It is difficult to see how such a view can be
maintained. For as the state of knowledge alters with time
different events come to be seen as genuinely requiring explanation and others pass out of the spotlight of attention.
Aristotle, for example, thought it necessary to explain what
kept a projectile in motion after it left the hand of the
projector but did not believe that any causal factor (other
than "nature") was needed to explain the circular motions
of the heavenly bodies.[25] Newton, on the other hand, was
convinced that the projectile's continued motion required
no special explanation but did believe that an account had
to be given of the forces holding the planets into their (approximately) circular paths.[26] Was Aristotle absolutely

---

[25] Aristotle, *De Caelo* I, 2: ". . . all movement that is in place,
all locomotion, as we term it, is either straight or circular or a
combination of these two, which are the only simple movements. And the reason of this is that these two, the straight
and the circular line, are the only simple magnitudes. . . . We
may take it that all movement is either natural or unnatural,
and that the movement which is unnatural to one body is
natural to another—as, for instance, is the case with the up-
ward and downward movements, which are natural and unnatural to fire and earth respectively. It necessarily follows that
circular movement, being unnatural to these bodies [fire, air,
earth and water], is the natural movement of some other [viz.,
the celestial aether]."

[26] Newton's *Principia,* Cajori's revision of the Motte translation (Berkeley: University of California Press, 1960), p. 13:
"Every body continues in its state of rest, or of uniform motion

wrong and Newton absolutely right? Who is to say? Peirce's
theory utterly fails to answer these questions while at the
same time implying that one of them was right and the
other wrong.

To compare the idea of "natural anomaly" to Peirce's
concept of "events requiring explanation" the following
diagrammatic representation will be helpful:

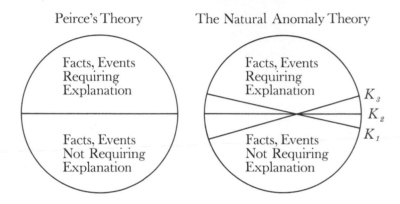

Peirce's Theory   The Natural Anomaly Theory

Each circle represents the totality of empirical facts. Peirce's
claim is that a sharp distinction can be made between the
two subclasses. I have maintained, rather, that every line
drawn between the two regions depends on the current
state of knowledge K. If the current state of knowledge
implies that event E ought not to occur, and if E does in
fact occur, then E requires explanation. If, on the other

---

in a right line, unless it is compelled to change that state by
forces impressed upon it." And (p. 2) "A body, from the inert
nature of matter [my italics], is not without difficulty put out
of its state of rest or motion."

hand, K implies that E should occur then we may say either that E has been explained or does not need to be explained. As our state of knowledge changes from say, K_1 to K_2 different facts or events become natural anomalies (as indicated in the diagram). At no time can we say without qualification, however, that an actual occurrence will *never* require explanation or will *always* require explanation.[27]

The concept of "the present state of knowledge" we have appealed to in the last few pages is, to be sure, a vague one. Taken in an unlimited sense it is probably self-contradictory: there is no such thing as a body of present knowledge acceptable to all reasonable men. The most one can plausibly require of explanation requests in ordinary discourse, therefore, is that they be contextually clear. It cannot be demanded that such requests be built around a naturally anomalous state of affairs. If someone asks us "Why do I have a nose on my face?" we are justified in asking "What makes you think you shouldn't?" In other words, we can properly ask for clarification about the beliefs which are presupposed by the original question. But it is a case of misplaced rigor to demand that a contradiction be shown between well-formed beliefs and presence of a nose on someone's face.

In the sciences, on the other hand, the situation is somewhat more definite. The idea of "the present state of knowledge" in, let us say, physics or chemistry has a

---

[27] A trivial exception to this is the case of tautologies like: "It is either Tuesday today or else it is not." An anomaly context for a tautology is always self-contradictory, hence tautologies never require explanation. This does not mean, of course, that no one ever asks for explanations of tautologies. It simply means that such requests cannot be rationally justified.

clearly definable meaning in terms of available empirical data, along with universally accepted experimental results and generalizations. These make up the backbone of what scientists refer to as "the literature." Against such a background, requests for explanations can always be rationally justified or rejected as unwarranted. This represents an important and fundamental difference between the orderly and structured character of science and the looser, more informal world of everyday discourse. One does not dismiss one's child's request for an explanation on the grounds that his question is ill-conceived. But scientists make such judgments about their own questions and those of colleagues every day. The very process of scientific research depends upon it.

Our account of natural anomalies, therefore, must be restricted to those scientific contexts where the notion of "present state of knowledge" has a clear and precise meaning. Outside that sphere it can serve as a tool in unearthing the presuppositions of a request for explanation but cannot ordinarily be used to rule out particular requests as illegitimate. The one major class of exceptions to this, are cases like "Why is there something rather than nothing?" when, as we previously noted, the presuppositions are themselves found to be inconsistent. Here it is clear that the question itself must be rejected as essentially irrational. But in general one would not be able to make such an assertion.

INTER- AND INTRATHEORETICAL EXPLANATION

Every scientific explanation involves, implicitly or explicitly, two sets of theoretical and observation statements, not just one. The first of these is the set of assumptions which logically entails that the explanandum-event or fact should not be the case. The second is the explanans which accounts for the actual state of affairs. Schematically, the situation appears this way:

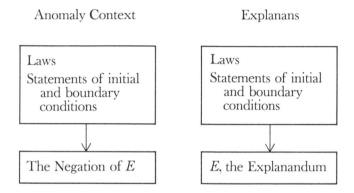

Anomaly Context   Explanans

Laws	Laws
Statements of initial and boundary conditions	Statements of initial and boundary conditions

| The Negation of E | E, the Explanandum |

The arrows here should be taken to represent deductive connection. According to restriction (R_6) imposed earlier the relation between the anomaly context and the explanans is an intimate one: nothing is to be altered in the anomaly context except what must be altered in order to square with the facts (and the logical conditions already imposed on the explanans). They are interlocked conceptually except in cases where our most fundamental assumptions must be abandoned or rearranged.

In this section we shall seek to draw out some implications of the intimate relationship between anomaly context and explanans in scientific explanation. These implications bear not only on the idea of anomalousness but also on such fundamental issues as the nature of law-like generalizations and scientific theories.

The semantical and logical relations obtaining between the anomaly context and the explanans are of decisive importance in providing the latter with its so-called "explanatory power." In fact, they play a key role in giving the laws of the explanans their law-like status. Granted there are a whole host of meanings for the term 'law,'[28] there is nevertheless one fundamental sense of the term which is best understood in terms of the model we are advancing. This sense is the one associated with such properties as functional *a priori*-ty, relative invulnerability to disconfirmation, capacity to support counter-to-fact conditionals, and so forth. *Such laws are marked in the present connection by their occurrence in both the anomaly context and the explanans for at least one—usually more than one—explanation.*

The principle of rectilinear propagation of light, for instance, is a law in this sense. It functions both in Newton's argument for the anomalousness of the elongation of the spectrum and in his explanation of it. It is, as it were, a constant factor in Newton's discussion. Yet this is not a mere matter of tenacity on Newton's part. He *did* check experimentally to be sure that the principle had not been falsified by the experiment. The principle was only retained because it was found to be true in the case at hand.

Similarly, Newton's law of universal gravitation and his laws of motion furnish a constant conceptual background for Leverrier's researches on Uranus. The alteration in assumptions which results in an explanation of the planet's behavior concerns primarily the boundary and initial conditions invoked rather than the laws themselves.

When a law has been found to survive logical crises of the kind here described it is a short step to the application

---

[28] Hanson, *Patterns of Discovery, op. cit.,* Chapter V.

of it in support of counterfactual inferences. Thus, the
law can be safely employed along with false assumptions
in order to tell us what *would* happen were those false
assumptions true. A case in point: "If there were no sheet
of paper here before my eyes, light from the lamp would
travel straight through the places where it now is reflected
back to my eyes." I have good grounds for using the principle of rectilinear propagation to draw this inference
since it has previously withstood tests in which it predicted
(a) what *should* happen under conditions which did not
actually obtain as well as (b) what *did* happen under actual
conditions. Accordingly, I expect the principle to make
the logical shift from "ought" to "does" equally well in
the present instance. This is one way, though certainly
not the only way, in which a generalization can come to
have the capacity to support counter-to-fact conditionals.

Much of the special "semantical content" which accrues
to law statements, then, derives from their success in playing a dual role: anomaly-identifier and anomaly-explainer.
In many cases, however, there is no such "bridging role"
to be played. What can we say about these instances? And
into what kinds of categories can explanantia be placed on
the basis of their peculiar makeup? Several options seem
to be possible.

One possibility is that the anomaly context and explanans may differ only in the boundary and initial condition statements they contain. This occurs, for example, in
Leverrier's explanation of the behavior of Uranus. The
boundary condition statement in the anomaly context
providing for the dynamical closure of the system "Jupiter-
Saturn-Uranus-Sun"—which is false unless stated as a
rough approximation—had to be replaced by a declaration
of the closure of the system "Jupiter-Saturn-Uranus-Sun-

Neptune." And the list of initial conditions had to be augmented by inclusion of the orbital elements of Neptune.

Typically, the explanans will be of this kind in cases where either (a) an observational error has been committed or (b) relevant (causal) variables have been omitted in the anomaly context.

A second possibility is that the anomaly context and explanans may differ in some, but not all, of the general laws they contain. Depending on the logical form of the laws this may or may not involve an alteration in initial and boundary conditions. When, for instance, Newton modifies Snell's law to account for the spread of the spectrum he must add mention of the fact that the incident beam consists of white light. Previously, it was not necessary to mention this. At any rate, Newton does retain other initial and boundary condition statements and preserves the rectilinear propagation principle as well.

The third and final possibility is that no general laws whatever may be preserved in the transition from anomaly context to explanans. This situation is the hallmark of radical innovation and the emergence of revolutionary new theories. It is the dominant form of explanation in what historian T. S. Kuhn has labeled "Revolutionary Science." [29] An example is the transition between classical celestial mechanics and the relativistic outlook of the 20th century. *All* of the laws of classical mechanics undergo basic revision in the new theory. As a result, the relativistic

---

[29] See Thomas S. Kuhn, *The Structure of Scientific Revolutions* (Chicago: University of Chicago Press, 1962), p. 91ff. Professor Kuhn's brief and somewhat cryptic discussion of anomalies (pp. 52–65) should also be consulted.

explanations of phenomena involve wholly different premise sets from those employed in classical theory. It is true, of course, that the classical laws of motion can be generated as "special cases" or "limited approximations" from the relativistic field equations. But our point is that in every instance restrictions and conditions must be added to the older laws. None is retained in its original form. By way of contrast we may note again that when Newton offers a revision of Snell's law he leaves the principle of rectilinear propagation of light intact. At least one basic law remains constant, therefore, as he moves from anomaly context to explanans. This does not happen when Einstein explains the advance in the perihelion of Mercury relativistically.

Einstein's explanation of the photoelectric effect *via* the quantum hypothesis is another example of radical innovation. The anomaly context in this case consists of the principles of Maxwellian radiation theory—especially the assumption of continuous energy transmission. Einstein's explanans not only does not use these principles, it formally and explicitly contradicts them. It is for this reason that the year 1905 marks a radical break with the tradition of Maxwell and Hertz.

On account of the persistence of a constant theoretical backdrop in the first two types of explanation discussed above, let us refer to them as *"intratheoretical* explanations." Explanations of the third type—e.g., the relativistic explanation of the behavior of Mercury or the quantum explanation of photoelectricity—will similarly be termed *"intertheoretical* explanations."

We are distinguishing here between explaining a fact *within* a particular theory and explaining it by means of a different or wholly novel theory. (Accordingly, we are also making a fundamental point about the nature of theories

at the same time.) One need only look again at the Hooke-
Newton controversy to see the importance of the distinction. Hooke sought to provide an *inter*theoretical explanation when Newton had already got hold of an *intra*theoretical one based on geometrical optics. Hooke's "anomaly context" was such as to imply that colors ought not to
be produced by mere refraction of light. He therefore
sought an explanation of the odd fact. Newton, on the
other hand, was thinking solely of the geometrical conditions relevant to his observations. The obvious moral of
this situation is that an intratheoretical explanation always
takes logical precedence over an intertheoretical one (other
things being equal). *The requirement* (R_6) *of contextual
relevance for explanation encapsulates this idea.* Deviation
from the context of present knowledge and belief must
always be minimal in scientific explanation; innovation for
innovation's sake is to be avoided. Science has, as it were,
a built-in conservatism. Thus, in spite of Einstein's great
success in explaining the photoelectric effect (*inter*theoretical explanation) we must acknowledge that it is the absence of a satisfactory intratheoretical explanation within
electromagnetic wave theory which gives logical respect-
ability to his results. An intertheoretical explanation over-
laid onto a situation where an adequate intratheoretical
explanation is possible is just so much excess baggage. This
point will be relevant in the sequel when we come to
discuss so-called "hidden variable" reconstructions of modern quantum mechanics in Chapter 5.

GENETIC AND LOGICAL ORDER

The confusion of logical structure of scientific theories
with their historical development is a cardinal example

of the well-known genetic fallacy. Yet it continues to be committed frequently by writers on scientific methodology. Accordingly, it is necessary for us to distinguish clearly here between the logical form of explanation as we have outlined it and the possible genetic and historical orders compatible with that form. This should serve to prevent a number of natural misunderstandings. It also will enable us to answer an important question posed in Chapter 1 concerning the methodological views of Einstein and Newton.

Which comes first, the hypothesis or the observation? The explanans or the explanandum?

Either. There is no proscription against a scientific discovery's taking either of the following routes:

Route I

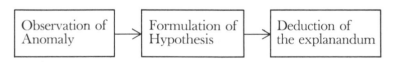

Route II

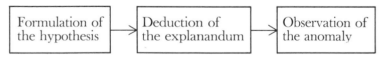

In point of fact, several other "routes" can be listed and examples given if we admit "Definition of the Anomaly" as a distinct temporal process involving a complex of activities. And when the possibilities for parallel and intermittent development of some of the processes are added, the list becomes very long indeed.

It turns out, then, that a multitude of genetic paths are

possible avenues to the birth of scientific explanations; so
many that there is little point in seeking to itemize them.
To speak, therefore, of *"the* method" of arriving at an
explanation is to misrepresent the actual situation. Explanations are constructed according to a logical map,
all right; but the direction of the scientific investigator's
inquiry and his starting point are not specified by the
map at all. He can begin with "experience" or with "concepts." The accidents of the situation determine which
method is preferable.

Newton and Einstein evidently believed that the logic
of scientific explanation included some kind of guidelines
vis-à-vis the direction, the starting point and the goal of
inquiry. Thus, Newton urges in his *Opticks* that

> . . . in Natural Philosophy, the Investigation of diffi-
> cult things by the Method of Analysis ought ever to
> precede the Method of Composition [Synthesis]. This
> analysis consists in making Experiments and Observa-
> tions, and in drawing general conclusions from them
> by Induction. . . . And the Synthesis consists in as-
> suming the causes discover'd, and establish'd as Prin-
> ciples, and by them explaining the Phaenomena pro-
> ceeding from them, and proving the Explanation.[30]

Einstein, however, registers a strong protest based on his
own experience with the theory of relativity.

> I have learned something . . . from the theory of grav-
> itation: No ever so inclusive collection of empirical

---

[30] Newton, *Opticks, op. cit.,* pp. 404–405.

facts can ever lead to the setting up of such complicated equations. *A theory can be tested by experience, but there is no way from experience to the setting up of a theory.* Equations of such complexity as are the equations of the gravitational field can be found only through the discovery of a logically simple mathematical condition which determines the equations completely or [at least] almost completely. Once one has those sufficiently strong formal conditions, one requires only little knowledge of facts for the setting up of a theory; in the case of the equations of gravitation it is the four-dimensionality and the symmetric tensor as expression for the structure of space which, together with the invariance concerning the continuous transformation-group, determine the equations almost completely.[31]

In all likelihood, Newton is speaking of explanation of *intra*theoretical type while Einstein has in mind the deep and searching overhaul of basic theory which so often marks *inter*theoretical explanation. But even if we assume them to be speaking of the same type of explanation it turns out that their methodological positions just represent two sides of the same coin.

Contra Einstein. It would be entirely possible for a person to encounter the observational and experimental evidence which now supports the special and general the-

---

[31] Einstein, in *Albert Einstein: Philosopher-Scientist, op. cit.,* Vol. 1, p. 89, my italics.

ories of relativity and to be led from this anomalous data[32] through a basic revision of the fundamentals of Newtonian theory to the relativistic field equations.

Contra Newton. Einstein did not actually discover his law of gravitation by generalizing or otherwise reasoning from experimental and observational data.

Pro Einstein. Newton probably didn't discover his gravitational equation by "deducing from the phenomena" either—the falling apple legend notwithstanding.

Pro Newton. If any general law stands in a determinate logical relationship to observational data then *whether we have a name for it or not* there is a converse logical relation between the data and the law. So if you can get (logically) from the law to the data—as Einstein concedes —then you can also get (logically) from the data to the law. It is not even a psychological impossibility.

The chief danger in partial and one-sided views of scientific explanation is that they harden easily into methodological dogmas. The history of Newton's views is well-known in this respect. In our own time the spread of an oversimplified version of the Einstein methodology has at times threatened to do similar damage to the delicate balance between theory and observation—this time at the expense of the empirical side. The obvious antidote to either of these excesses is a clear understanding of the logical structure and factual bases of our theories.

---

[32] Anomalous with respect to Newtonian mechanics, of course.

3

Anomalies and Theories

Euclid provided the model of a deductive, demonstrative science when he gathered Greek geometry into his *Elements*.[1] Physicists and philosophers have sought ever since to fit physical theories into the same logical mold. Only occasionally has this enterprise been questioned.

As Euclid depicts it, geometry is organized around a set of fundamental propositions (the axioms and/or postulates) and everything else is deduced—literally deduced—from these. Definitions serve as tools either for clarification of the intended semantical content of the abstract concepts or as notational simplifications intended to introduce more perspicuous ways of speaking. From the postulates one infers theorems and their corollaries via strict reasoning and, in principle, without reliance on diagrams, pictures or other visual aids. The system is conceptually closed and purely formal—though, for heuristic purposes, it is useful to think of it as having a definite semantical content or physical meaning.

---

[1] *Euclid's Elements,* tr. by Sir Thomas L. Heath, second edition, three volumes (New York: Dover Publications, 1956).

Are physical theories anything like the theories of pure mathematics? In certain respects the answer is surely "yes." In others there is room for doubt. The main thing is that physical theories explain things about the workings of Nature while theories of pure mathematics do not. Nevertheless, it must be conceded that physical theories do rest upon axiomatic foundations in most cases and that there is little logical difference between the kinds of arguments and inferences used by the physical scientist and those used by the mathematician.

The standard logical analysis of theories as axiomatic systems is known as the Hypothetico-Deductive theory of theories. It is a natural and plausible extension of the Deductive-Nomological pattern of explanation if we assume that theories are comprehensive systems of scientific explanations. One of the most persuasive of the Hypothetico-Deductive logicians, R. B. Braithwaite, puts it this way:

> A scientific system [theory] consists of a set of hypotheses which form a deductive system; that is, which is arranged in such a way that from some of the hypotheses as premisses all the other hypotheses logically follow.[2]

In other words, hypotheses of greater or less generality take the place of the axioms or postulates of a pure mathematical theory but the logical structure of the theory is essentially the same. *The hypotheses of the theory, if true, may come to be given the honorific title of 'law.'* The procedure by which this comes to pass—the procedure of con-

---

[2] R. B. Braithwaite, *Scientific Explanation, op. cit.,* p. 12.

firmation—consists essentially in checking deductions from the theory (predictions or explananda) against empirical facts. As the theory piles up successful predictions or "confirmation instances" its probability increases until, at some point, we begin to speak of the constituent hypotheses as laws.

The 'deductive' in 'hypothetico-deductive' must be understood in the technical sense of the term; i.e., as either syllogistic or mathematical derivation of one proposition or formula from one or more other propositions or formulae. Because the word 'deduction' has often been used with less precise denotation this point needs to be borne in mind. The connections between hypotheses are *purely* formal in a hypothetico-deductive type theory. They can, in principle, be carried out by a properly programmed computing device.

Braithwaite and other H–D theorists use the fact of deductive connection between hypotheses in theories to give a definition of the concept of "level" within a scientific system.

> The propositions in a deductive system may be considered as being arranged in an order of levels, the hypotheses at the highest level being those which occur only as premisses in the system, those at the lowest level being those which occur only as conclusions in the system, and those at intermediate levels being those which occur as conclusions of deductions from higher-level hypotheses and which serve as premisses for deductions of lower-level hypotheses.[3]

---

[3] *Ibid.*, p. 12.

Paralleling the levels of hypotheses, according to Braithwaite, are levels of generality. Accordingly, a highest-level
hypothesis may very well include terms which have no
direct experiential or observational reference.

> As the hierarchy of hypotheses of increasing general-
> ity rises, the concepts with which the hypotheses are
> concerned cease to be properties of things which are
> directly observable, and instead become 'theoretical'
> concepts . . . which are connected to the observable
> facts by complicated logical relationships.[4]

The semantical content of the theory, then, depends ultimately upon the lower-level hypotheses of the theory—
the hypotheses which describe observable conditions. The
higher-level propositions may turn out to be at least partially uninterpreted empirically. (By contrast, a theory in
pure mathematics can always be thought of as *wholly* uninterpreted.)

The hypotheses which go to make up a scientific system,
the H–D theory affirms, are linked either by mathematical
or syllogistic rules of logical inference. And depending on
the relation of deducibility, the hypotheses separate into
classes according to distinct levels of generality. Braithwaite, in fact, goes further: not only must *some* of the
hypotheses in a scientific system be so linked to others, *all*
hypotheses must be so linked.[5] This is almost implied by
the definition of levels mentioned above; for any hypothesis
which is deductively independent of all other propositions

---

4 *Ibid.,* p. vii.
5 *Ibid.,* pp. 18–19.

in the system will have no level and, consequently, will not be part of the system.

The full significance of the H–D theorists' commitment to a deductive interconnectedness of the hypotheses in scientific theories can only be fully appreciated when we come to consider the role of falsification and anomalies in the H–D system. Here the idea of "testability of hypotheses" is given concrete elaboration and the actual interplay between theory and observation is confronted.

The empirical testing of the deductive system is effected by testing the lowest-level-hypotheses in the system. The confirmation or refutation of these is the criterion by which the truth of all the hypotheses in the system is tested. The establishment of a system as a set of true propositions depends upon the establishment of its lowest-level hypotheses.[6]

A theory, to use Professor Willard Quine's happy metaphor, meets experience only at its periphery.[7] But a successful or unsuccessful test experience there can have repercussions reaching far into the interior. A confirmed

---

[6] *Ibid.*, p. 13.

[7] W. V. O. Quine, *From a Logical Point of View* (Cambridge, Mass.: Harvard University Press, 1953), p. 42 ("Two Dogmas of Empiricism"): ". . . total science is like a field of force whose boundary conditions are experience. A conflict with experience at the periphery occasions readjustments in the interior of the field. Truth values have to be redistributed over some of our statements. Reëvaluation of some statements entails reëvaluation of others, because of their logical interconnections . . ."

prediction derived from the theory serves to confirm the whole theory in some measure. And a false prediction signals the existence of some difficulty within the set of hypotheses from which the prediction was first deduced.

The logic of the falsification procedure can be put even more precisely according to the H–D theory. If hypotheses $H_1, H_2, \ldots, H_n$ are employed in deducing the observational consequence H and if H turns out experimentally or observationally to be false, then as "a simple matter of deductive logic" at least one of the hypotheses H_i must be regarded as false. If that H_i is, in turn, a consequence of higher-level hypotheses $H'_1, H'_2, \ldots, H'_m$, then at least one of these must be false as well. And so on. It is in this way, H–D theorists maintain, that anomalous experiences can reach into the center of our field of knowledge and force changes and reassessments.

The logician's phrase 'at least one' used in the preceding paragraph introduces an element of vagueness into the H–D account of anomalousness and falsification of theories which cannot be ignored. Braithwaite seeks to clarify the matter this way:

> . . . suppose, as is frequently the case, that we are considering a deductive system in which there is no one higher-level hypothesis from which (the falsified) lowest-level hypothesis follows, but instead the system is such that this follows from two or more higher-level hypotheses. Then what will be refuted by the refutation of the lowest-level hypothesis will be the conjunction of these two or more higher-level hypotheses: what will be a logical consequence of the falsity of the lowest-level hypothesis will be that at least one of the higher-level hypotheses is false. Thus in the case of almost all scien-

tific hypotheses, except the straightforward generalizations of observable facts which serve as the lowest-level hypotheses in the deductive system, complete refutation is no more possible than is complete proof. What experience can tell us is that there is something wrong somewhere in the system; but we can make our choice as to which part of the system we consider to be at fault. In almost every system it is possible to maintain any one hypothesis in the face of apparently contrary evidence at the expense of modifying the others. . . . But at some time a point is reached at which the modifications in a system required to save a hypothesis become more unplausible than the rejection of the hypothesis; and then the hypothesis is rejected.[8]

On the other hand, even when a theory is known to be falsified, the absence of a suitable alternative may force scientists to retain it for some time.

Long before Einstein propounded his theory of gravitation it was known that Newton's theory could not account by itself for the observed motion of Mercury's perihelion. But Newton's theory was not dethroned until Einstein's theory was available to take its place. The process of refuting a scientific hypothesis is thus more complicated than it appears to be at first sight.[9]

However complicated, the process is not logically obscure, though, and Braithwaite's final word on the subject speci-

---

[8] Braithwaite, *op. cit.,* pp. 19–20.

[9] *Ibid.,* p. 20.

fies the conditions for falsification with complete lucidity. First, he entertains the following objection:

It is simply not the case that every universal hypothesis is regarded as having to be rejected by the discovery of one contrary instance; the propounder of the hypothesis may well say that he did not expect it to hold under all circumstances, and that the fact that it does not hold universally does not show that it is not a good approximation to the truth.

The reply is then as follows:

[There are] two cases in which a hypothesis will not be regarded as definitely refuted by contrary instances. The first case is that in which the hypothesis is a statistical hypothesis. . . . The second case is that in which the hypothesis, though not explicitly of the form of a statistical hypothesis, will be treated as a statistical hypothesis in that it is to be rejected (and only provisionally rejected at that) only if the contrary instances show deviations from the value asserted in the hypothesis which exceed a certain amount. If the deviations are less than this amount, the instances will be taken as being affected by 'errors of observation' rather than as refuting the hypothesis.[10]

In the absence of errors of observation or statistical hypotheses, therefore, the falsifying process is precisely as has been previously described: a simple matter of deductive logic.

---

[10] *Ibid.*, pp. 360–61.

DIFFICULTIES IN THE H–D ACCOUNT

Taking Braithwaite's treatment as a guide we can summarize the H–D account of the nature of non-statistical scientific theories rather briefly with the aid of a diagram. What the H–D account says, in essence, is that a scientific theory is a body of hypotheses (putative laws) which when supplemented by physical data statements—initial and antecedent condition statements—operates as a pure deductive axiom system. In other words, theories are just axiom systems of the same type as mathematical axiom systems except for the division of axioms into two classes: hypotheses and data statements. We schematize this as follows:

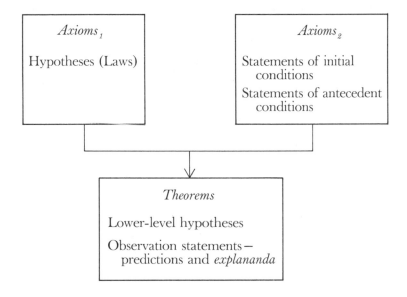

There are a number of difficulties with this interpretation of scientific theories. Some of these have been conceded by Braithwaite and other H–D theorists but others have not. In this section I should like to consider in some detail the defects in the H–D account, both acknowledged and unacknowledged. The full analysis of some of these difficulties will carry us on in succeeding sections to more detailed examination of the nature of theories.

Defects in the analysis of law-like generalizations. Braithwaite allows that the analysis of the concept of lawfulness he is able to offer using the idea of scientific Hypothetico-Deductive System is not quite a "knock-out blow" to the opposition.[11] The attempt to explicate such generalizations as 'All light travels in straight lines' as general hypotheses in a system of deductively-linked hypotheses makes it difficult to explain the vast semantical differences between such laws and "mere" generalizations —like 'All ravens are black.'

If we contrast the H–D treatment of the idea of law with our discussion of anomalies in Chapter 2 it appears that the missing ingredient is the lack of any consideration of the complex logical relations between the law-like generalization, its fellow generalizations in the theory and its empirical consequences. *According to the H–D account, the only logical relation between hypotheses within a theory is that of deducibility.* One of the hypotheses will be a consequence of the other or else they will be logically independent within the theory. They may have mutual consequences but this does not affect their basic independence.

As all physicists recognize, this is not an accurate picture of the relations between the fundamental laws of a physical theory. Semantically, at least, the basic laws are inter-

---

[11] *Ibid.,* p. 317.

related in any well-structured physical theory. The H–D
analysis gives us no hint as to how this can be. Our prior
discussion of anomalies does. For from that discussion it
would appear that the role of a law, like the principle of
rectilinear propagation in intratheoretical explanation, sets
up a special relation between it and such other generalizations as Snell's law.

Snell's law is not deducible from the principle of rectilinear propagation. In a strictly H–D theory these two
laws would be construed as logically independent. But the
persistence of the rectilinear propagation principle in such
explanations as Newton gave of the elongation of the spectrum gives it a special significance with respect to Snell's
law. *Because it functions as a constant theoretical background for Newton's investigation of the limitations of
Snell's law the principle of rectilinear propagation takes
on the status of a "higher-order" generalization.* We are
led to expect that it will survive future crises even if Snell's
law should have to be abandoned.

The modification which is most needed in the H–D
account of law-like generalizations is the inclusion of some
reference to the intratheoretical explanatory role of such
generalizations. For it is this role which gives them their
special resiliency in the face of apparent counterevidence
and which welds a motley collection of disparate general
statements into a unified body of knowledge.

Experimental generalization. Another defect in the
H–D account—this one evidently not acknowledged by
Braithwaite—is the absence of any treatment of the role
of theoretical assumptions in what Newton used to call
"deduction from the phenomena." It appears from the
H–D account that all scientific inference must proceed
either (i) deductively from hypotheses to observations or

(ii) inductively and without benefit of any theoretical background from observations to lower-level generalizations. We are left with the impression that the discovery and, more importantly, the logical justification of higher-level generalizations must be a most mysterious and mystical affair.

In point of fact, there is a third type of inference which can be made and which is the stock in trade of the experimental physicist. Newton's phrase aptly describes this kind of inference. For it is just that: a deductive inference from observational or experimental data to a theoretical claim. Naturally, this kind of inference does not proceed without the invocation of some more or less general physical principle.[12] You cannot literally deduce a general statement from a class of individual or particular statements. But the inference does not rely on anything so metaphysical and broad as the Principle of the Uniformity of Nature.

Newton's own experimental inference is a case in point. Presumably, his results are to apply to all light anywhere

---

[12] Compare with Stephen Toulmin, *The Philosophy of Science, op. cit.,* pp. 40–41: "[Ernst Mach] thought that we should be justified in accepting our theoretical conclusions only if these were logically constructed out of the reports of our experiments; that is, related to them in a deductive way, as strictly as statements about 'the average Englishman' and data about individual Englishmen. . . . The confusion of thought which led Mach and the Phenomenalist School to this conclusion is not entirely easy to sort out. . . . But it is essential to see at the outset that there can be no question of observation-reports and theoretical doctrines being connected in the way Mach thought: the logical relation between them cannot be a deductive one."

and to any well-made prism. Yet he does only one experiment and repeats it at most a few times. How, as an inductive inference, can this be tolerated? How can we count his version of Snell's law as well-confirmed when such a limited number of "confirmation instances" have been realized? The answer is that we have good theoretical reasons to assume independently of this particular *experimentum crucis* that all light has the same basic properties. With the exception of its position, orientation, wavelength, polarization and few other properties, one light ray is the same as any other. This latter generalization is based not on some imaginary Uniformity of Nature but on the *de facto* regularity of light beams as previously examined in nature. And it, in turn, provides the logical basis for Newton's general conclusion that all white light will contain differently refrangible components.

In modern particle physics, the essential identity of all electrons provides a similar basis for deductive generalization. One well-performed experiment can provide all the evidence needed for a general conclusion about all electrons.

The H–D account of confirmation ignores this third type of inference entirely. As a result it hits far wide of the mark in its description of the way in which empirical support for a theory piles up. Nor is this a shortcoming in the H–D account alone. Almost the entire branch of philosophy of science known as "inductive logic" has been founded on the premise that in a scientist's generalization from experiments and observations the *modus operandi* is little different from a blindfolded man's drawing black and white balls from an urn.

Proponents of the H–D model may object at this point that our discussion of experimental generalization rests on a confusion between the contexts of discovery and justification. The meaning of such a charge, I take it, is that although generalizations may be *discovered* by utilizing a background of theory and inferring from empirical data, nevertheless a generalization cannot be considered to be *justified* by that kind of inference.

On the contrary: scientific generalizations can be and are justified by precisely the kind of deductive argument we have been discussing. The error involved in the H–D theorist's objection is the error of assuming that since empirical or observational statements represent the only source of confirmation for a theory taken as a whole they must also be regarded as the only source of confirmation for individual propositions of the theory. The theory θ, on this view, is regarded as being confirmed by the evidence statements e_1, e_2, . . . , e_n and it is then inferred that any law-like statement L_1 included in θ must derive *all* of its evidential support from the e_i and the e_i alone. This is a non sequitur. For if another law-like statement L_2 in θ is such that L_2 and some of the e_i deductively entail L_1 then L_2 is part of the evidence for L_1. To recall an example alluded to a moment ago—part of our reason for believing that all electrons obey the Heisenberg uncertainty relations is the quantum mechanical postulate that there are no differences among electrons which would enable us to distinguish them (save position, momentum, etc.). An electron which did not obey the uncertainty relations would surely have such a differentiating property. Obviously, this logical relationship between the basic laws of the quantum theory has nothing to do with the order in which the Heisenberg relations or the identity postulate were discovered.

The logic of generalization in physics will be examined further in Chapter 4. Suffice it to say here that the usual

presentations of the H–D account give insufficient attention to the mutually supportive roles that generalizations
may play within a given theory.

   Scope and boundary conditions.   Still another omission
in the H–D account is the lack of any detailed consideration of the special role of boundary conditions in the formulation of scientific laws. By 'boundary conditions' here
I do not mean only those numerical conditions used to
specify the state of a boundary in thermodynamics, fluid
mechanics, etc. (though these are certainly to be included).
What I have in mind is the totality of conditions attached
to a law (or laws) and intended to mark out the sphere of
its applicability; that is to say, the set of conditions which
describe an appropriate system of bodies or objects for the
application of the law.

   Stephen Toulmin has suggested the title of 'scope' for
such a list of conditions. As Toulmin points out,

> Any one branch of physics, and more particularly any
> one theory or law, has only a limited scope: that is to
> say, only a limited range of phenomena can be ex-
> plained using that theory, and a great deal of what a
> physicist must learn in the course of his training is
> concerned with the scopes of different theories and
> laws. It always has to be remembered that the scope
> of a law or principle is not itself written into it, but
> is something which is learnt by scientists in coming
> to understand the theory in which it figures. Indeed,
> this scope is something which further research is always
> liable to, and continually does modify . . .[13]

---

[13] *Ibid.,* p. 31.

Reviewers have pointed out that Toulmin's assumption that scope and law are always separated in the statement of a theory is not strictly correct. Scientists would surely agree that it is generally advisable (though practically awkward) to include both in any specification of the theory.

Even assuming that scope and law are jointly stated in some way, though, it is far from evident how the H–D theorist is supposed to take account of scope in his scheme of things. Where exactly should a specification of scope come in?

A plausible answer—and the most natural way of extending the H–D account—is to include scope conditions as part of the antecedents listed in the hypothetical form of the law. That is, we should place the scope conditions in the "if-part" of the law. In this way, a law of the form 'If $A, B, \ldots C$, then E' having scope limitation of the form 'In a system with properties $P, Q, \ldots, R$' will be transformed into a statement of the form 'If $A, B, \ldots C$ and $P, Q, \ldots, R$ obtain, then E.'

The idea of scope, however, is an extremely slippery conception. (Toulmin, who deals as thoroughly with it as almost any contemporary writer, leaves the term essentially undefined!) We shall therefore have to consider this proposed extension of the H–D account carefully in a later section before regarding it as wholly satisfactory. In fact what we shall find is that it is not really a defensible account of scope in physical theories at all.

Idealizations. The H–D model does include provision for so-called theoretical terms—terms referring to unobservables. But it is not clear that this treatment is also intended to apply to a more basic kind of strategy common to many physical theories: the idealization of conditions.

'Idealization' here does not mean merely the employment of mathematical functions and relations (especially geometrical) to represent physical phenomena. It refers rather to *the use of scope or boundary conditions which are never realized in nature.* Archimedes' demonstration of the law of the lever, for instance, is supposed to hold for an "ideal" beam balance.[14] Newton's law of inertia is supposed to apply to "dynamically closed systems of single bodies." The Boyle-Charles gas law is supposed to apply to an "ideal gas." None of these things is ever found in nature. Is that what the H–D theorist means when he speaks of a theoretical term? Or is he only dealing with concepts like electron spin, molecular bonding force, etc., —things not directly observable but actually existing in nature? The difference is obviously an important one.

The absence of any account of scope conditions in the H–D doctrine leads to the conclusion that it does not purport to deal with idealizations of this kind. For the hallmark of such idealizations, again, is that they involve the assumption of ideal and empirically non-realizable boundary (scope) conditions for certain laws.

Approximation. Closely related to the failure of the H–D account to deal with scope conditions and idealizations is the complete absence of any method for dealing with approximative inference. The tacit assumption of the H–D theorist is that all scientific inference begins with full data about the empirical situation and deduces predictions and *explananda* from that data. But in practice scientists rarely proceed in this fashion. Almost invariably,

---

[14] See *A Source Book in Greek Science,* ed. M. R. Cohen and I. E. Drabkin (Cambridge: Harvard University Press, 1958), pp. 186–89.

predictive inferences are based on less than complete lists
of initial and boundary conditions. How is this possible
on the H–D view?

An H–D theorist may counter at this point by arguing
that we have raised a pragmatic, rather than a logical point.
But that remains to be seen. *For if idealizations play the
role we have so far suggested it follows that for inferences
in some theories "complete" empirical data may not exist.*
This will be explained more fully below. In any case, it is
worthwhile even as a pragmatic point to ask how approximative inference takes place. For the analysis of *de facto*
scientific practice can go a long way to enhance our understanding of the limitations and applicability of a logical
scheme like the H–D pattern. And more importantly, it
can go a long way toward enhancing our understanding of
science itself.

Falsification. Finally, as a synoptic way of stating our
previous objections to the H–D model, we can say that the
logic of falsification of theories implied by the H–D account is simply not commensurable with actual scientific
practice.[15] Nor is it internally satisfactory, as Braithwaite's
own remarks have already suggested. The idea that a single

---

[15] The appeal to scientific practice as a test for the reliability
of a logical account has recently come under fire from the
arsenal of Robert Ackermann (Cf. his review of Mary Hesse's
*Models and Analogies in Science. British Journal for the Phi-
losophy of Science,* Vol. 41 (1965), p. 161.) According to Ackermann, "There can be little doubt that most of the important
contemporary discoveries in philosophy of science have resulted
from the consequences of a distinction between the context of
discovery and the context of justification and a subsequent
application of the techniques of symbolic logic to problems

prediction's coming out false should be logically sufficient to overthrow a theory is simply ludicrous when measured against the history of science. *The occurrence of a false prediction will normally lead scientists into that type of investigation we previously labeled as "definition of the anomaly."* Rarely will they throw up their hands in dismay and admit that a cherished theory has been dethroned.

This is not a matter of psychological obstinacy. It is

---

which can be formulated in the context of justification. . . . It has been objected that this approach is not *descriptive* of scientific practice since fully axiomatised theories are rarely used by scientists, and further that this approach is not capable of furnishing any answer to the question of why certain formalised theories rather than others should come to be adopted. These objections are often meant as criticisms of the formalistic approach, but such criticism is misplaced in so far as formalists only claim to be attacking a specific range of problems." It is difficult to see, however, how this excuses the formalists from the obligation of squaring their work with actual scientific practice. To suggest an analogy: suppose one were purporting to give an axiomatization of the first-order predicate calculus but the set of axioms one chose did not allow the Deduction Theorem to be proved of the axiomatization. Could one then excuse oneself from criticism by saying: "I only claimed to be dealing with a limited and specific range of problems and I am therefore under no obligation to square my system with the practice of working logicians and mathematicians"? Or would one rather be obliged to label one's system as a *special type* of predicate calculus? Perhaps what the formalists need is such a special label for their discipline—one which does not suggest that they intend to convey understanding of the actual workings of the natural and social sciences. Surely that is what is conveyed in the title 'Philosophy of Science.'

something more like a case of rational resistance based on lack of evidence. In other words, the logical resources of a well-made physical theory are such as to require strong empirical reasons for its abandonment. A single false prediction does not normally provide that kind of strong reasons. For a fuller explanation of why this should be so it will be necessary for us to probe more forcefully into the ideas of scope, approximation and idealization.

APPROXIMATIVE INFERENCE

An approximative inference in a physical theory is, by definition, one in which observation statements are deduced without full knowledge of the scope conditions affecting the system we are dealing with.[16] It is an inference designed to be more or less correct but which is not normally expected to be exactly "right."

Now the question is: how can the H–D model of scientific theories be expanded or modified so as to take account of such inferences?

The natural starting point is to assume that scope con-

---

[16] This definition of 'approximative inference' should be compared with C. G. Hempel's discussion of what he terms "approximative explanation." (Cf. *Aspects of Scientific Explanation, op. cit.,* p. 344.) Hempel's approximative explanations are explanations of *laws*—rather than observations. The paradigm example is the explanation of Galileo's law of falling bodies as an approximation within Newtonian classical mechanics. Since this is relatively straightforward deductive explanation of a slightly modified version of Galileo's original law it is not altogether clear why Professor Hempel attaches a special label to it.

ditions are parts of the antecedents of the laws themselves. In other words, pack the scope conditions in among the initial conditions and treat them in a logically analogous fashion. Specifically, this means: *treat approximation in the same way that observational or experimental error in initial conditions is handled.* Thus, if a scientist assumes that boundary conditions of his laws are fulfilled (when in fact they are not) we shall regard the situation as one in which a certain "error" has been introduced into the initial-condition data. Presumably, a numerical value will be assigned to the error and it will be carried through all of our calculations in the usual way. Our predictions will come out with a certain "plus-or-minus" value indicating the limits within which they are to hold. This plus-or-minus value will simply augment the usual error value introduced by errors of observation in the measurement of the initial conditions. Here, then, is a direct and quantitative way of dealing with approximation.

In certain areas of physics, this approach would seem to have some validity. For instance, if in a thermodynamic calculation we employ a law applicable within an adiabatically bounded system to a system which is not quite adiabatically bounded (one which allows a small amount of heat to pass into and out of the system) a quantitative estimate of the "leakiness" of the walls can be introduced and used to calculate uncertainties in predictions generated from the law.

When we attempt, however, to carry through a close logical analysis of this procedure it turns out that things are not what they seem. The use of "limits of uncertainty" to characterize approximative inference bears almost no resemblance to the use of limits of uncertainty to deal with observational error. What is more, for a large class of physical cases in which approximations are used it is wholly impractical to speak of "calculating the uncertainty in the predictions."

To focus more clearly on the problem, let us take a reasonably complete formulation of Snell's law with scope conditions included in the statement of the law:

> *Snell's law:* If a beam of monochromatic light passes successively through two optically homogeneous, transparent and non-double-refracting media, then
>
> $$\frac{\text{Sin } i}{\text{Sin } r} = \text{constant},$$
>
> where i is the angle of incidence of the beam upon the second medium and r is the angle of refraction in that medium.

In this formulation, the boundary conditions on the beam and media include the restrictions to homogeneous, transparent and non-double-refracting media and monochromatic beams.

Suppose now that these conditions are only approximately met in a particular experimental situation. The beam is not purely monochromatic but includes light of several discernible wavelengths. The media are not wholly transparent but are highly translucent. One of them—say air—is not fully homogeneous but has currents and whorls in it. And so forth.

How should we assign numbers to these approximations? More importantly, how shall we assign limits of approximative error which can be translated meaningfully into terms of the expected certainty in the value of r? *The analogy to observational error breaks down right here.* The basic problem is that in the case of boundary conditions we often are not dealing with simple, operationally-defined quantities whose limits of uncertainty can be encapsulated in a "plus-or-minus" expression. There is no simple way to assign "limits of approximation" to something like the homogeneity of a medium. Homogeneity does not have a direct numerical index, nor does it enter into Snell's law in a quantitative way.

Suppose, however, that this objection could be surmounted. Suppose, for instance, that a numerical scale could be devised on which monochromatic light has an index of zero and "mixed" beams receive successively larger positive values according as they deviate "further" from being monochromatic. Suppose finally that in a particular experimental setup we observe that the light actually being used is of type n. We therefore list as a boundary condition of the experiment 'Beam of light is monochromatic—plus or minus n.' What then? How can the number n be meaningfully related to predictions of the value of r? Certainly not by the kind of simple calculations dealt with in the ordinary theory of experimental error! The angle r is mathematically related to the angle of incidence and the refractive constant by Snell's law. But the law does not, by itself, provide a mathematical relationship between the homogeneity of the media, the monochromaticity of the light, etc., and the angle of refraction. Such relationships, if they can be provided at all, must be provided by other optical laws.

Proponents of the H–D account will urge at this point that the handling of the boundary conditions is greatly facilitated from a logical standpoint if we include in our *explanantia* those "other optical laws" which relate homogeneity, etc., to the angle of refraction. And in the case of

Snell's law it is at least conceivable that this could be done. Then, of course, we should be able to assign appropriate uncertainties to the various initial conditions for these laws and show precisely by how much calculations using Snell's law can be expected to deviate from the actual observations. Measurements on the homogeneity of the media, wavelengths of the light, etc., would enable us to tell exactly how good an approximation a straightforward Snell's law computation gives in the present experiment.

Certainly, inclusion of the "other optical laws" will do all of these things. It will reduce approximateness in boundary conditions to observational error in initial conditions. But then we are no longer dealing with an approximative calculation at all! We are dealing with a highly complex, exact inference as a proxy for the work-a-day physicist's approximative inference! Our logical account takes on all of the grotesque unreality which scientists traditionally have ascribed to the philosophy of science.

Still, as a logical account of approximative inference with theories like geometrical optics the H–D account cannot at this point be faulted. The boundary conditions of such a theory are, in the long run, operationally definable and functionally related to the dependent variables of the laws. The mere historical fact that the theory has never been fully developed in this direction, that exact ways of computing effects of inhomogeneity of media on refraction were inaccessible to Descartes, Newton, Snellius and others —such facts have no real bearing on the abstract logical analysis of geometrical optics.

To sum up: the natural thrust of the H–D approach is to recast such laws as Snell's in a form where all boundary conditions are effectively eliminated. Or, what comes to the same thing, all of the boundary conditions are defined operationally and functional relations between the boundary conditions and dependent variables are established empirically. Then any computation involving approximation in the boundary conditions can be transformed into a computation involving only numerically assignable uncertainties attached to a rather long list of initial conditions.

In order to achieve a more realistic view of the nature of approximative inference let us begin by defining clearly the two different ways of formulating a physical law statement. On the one hand we have the *exact form* of the law. This is the form of statement in which all scope or boundary conditions are mentioned as part of the antecedent of the law. It is the form which, from the H–D standpoint, is basic.

On the other side, we have a whole range of possible formulations which we shall designate as the *approximative forms* of the law. These are forms in which one or more of the boundary conditions is suppressed or omitted. In the limiting case we obtain the *simple form* of the law, which merely sets out a relationship between initial (or antecedent) conditions and consequences. As examples of simple forms we might offer: 'All light moves along straight-line paths.' 'All planets move in elliptical orbits with the sun at one focus' (Scope: in a dynamically-closed two-body universe.). 'Action is always equal to reaction' (Scope: in a dynamically closed system of two perfectly elastic punctiform bodies), and so forth.

Approximative inferences can most conveniently be viewed as deductive inferences using approximative forms of physical laws. Thus, instead of explicit inclusion of boundary condition statements which are either in the strict sense false or, at least, not known to be true, we are proposing that approximative inference be viewed as inference based on partially stated laws.

From the standpoint of pure logic there is no difference. Using a false boundary condition statement or using a truncated law statement in the end amounts to the same thing. In practice, however, the boundary conditions are rarely mentioned if nothing is known about them so the "approximative form of the law" approach seems more natural.

Now the immediate objections to this characterization of approximative inferences can be put in something like the following form:

> Surely this cannot be the whole of the story. For in the first place, a deductive inference drawn from an approximative form of law is based on a statement which is known to be literally false! If scope conditions are omitted, the law-statement—even if believed to be true in the exact form—no longer holds true. So there is no guarantee that predictions inferred from the approximative form will be true. Secondly, there is not even a guarantee that predictions inferred from the approximative form will be *close* to the truth. If an approximative form is applied to a system for which the boundary conditions of the law do not even represent a rough description our predictions will be grossly in error. How, then, can they be spoken of as "approximations"?

The answer to this barrage of objections is fairly simple: there *is* no guarantee that an approximative inference will yield true predictions or even "close" ones. But there does not have to be. Approximative inference as such is not concerned with giving *good* approximations. To assure that an approximative inference yields a *good* prediction it is necessary to have external empirical evidence.

There are basically four types of external evidence that can be adduced to support an approximative inference. Some of them can, as a matter of practice, be supplied when predictions are made. *But for the most part predictive approximative inferences proceed with their eyes closed.* Only after our prediction comes out badly can we ascertain how good an approximation we ought to have got. The four types of justification include: (1) direct empirical testing of the limits of error introduced into predictions by variation of boundary conditions; (2) repetition of the experiment or observation under widely varied circumstances with the assumption being that this effectively varies boundary conditions; (3) rough computation of the limits of error based on the "greatest apparent source" of violation of boundary conditions; (4) pre-computation of the exact limits of error induced in the prediction by uncertainties in the boundary conditions. The last of these has already been discussed in connection with the attempted extension of the H–D approach above. Let us consider the other three individually.

Direct empirical test.   After a prediction based on approximative inference has failed to come close to the observed results it is often a simple matter to locate the cause of the difficulty by varying empirical conditions suitably. Thus, when Newton observed the elongation of the spectrum he was led to ask whether a flaw or inhomogeneity in the glass might be responsible. This can easily be checked by using different parts of the prism to pass the light and observing the effect on the image cast. In effect, what one does is to perform an optical test for homogeneity of the glass.

If we are thinking in terms of exact prediction here, of course, the empirical test method will be of little use in establishing the limits of error introduced by inhomogeneity. For in order to make an exact prediction one would probably have to resort to some kind of chemical or non-optical criterion in order to be absolutely sure of the size and nature of the inhomogeneities. The empirical test gives only a very rough index and cannot be regarded as a certain way of establishing uncertainties. To put it a bit bizarrely: uncertainties established by rough empirical test are always themselves slightly uncertain.

Rough computation. This method, again, lends itself better to *ex post facto* estimation of the uncertainty in a prediction. But it also has a good deal of utility prior to the formulation of a prediction. The technique can be illustrated easily in celestial mechanics where the use of two-body equations for short-range predictions of the behavior of Mars can be justified by reference to a quick calculation of the simple gravitational action on Mars of the nearest and largest body during the period for which the predictions are to hold. This gives a rough estimate of the maximum distortion of Mars' orbit during the period of the prediction and hence of the maximum error to be expected in longitude and latitude.

Repetition of experiment. Where neither of the foregoing methods applies, a technique often used is that of repeating the experiment or observation for which a prediction has failed. This can be assumed, if done in a sufficiently varied number of settings, to induce random variations in the boundary conditions, thereby giving the effect of a direct experimental variation of those conditions. This method, of course, is virtually useless in fixing the limits of error in a prediction before the prediction has been made and falsified.

The importance of these three methods for justifying approximative inference is perhaps most evident when we consider the testing of laws and theories. According to the H–D approach, any such test must employ full data, seemingly true hypotheses and deductive reasoning. But using the three methods of justifying approximative inferences *it is possible to falsify a theory without ever having the full data required by the H–D model.* A prediction based on an approximate form of a law may hit far wide of the mark. If we then follow up this failure with an attempt at justifying the approximative inference we may find that it cannot be justified! Variation of the appropriate boundary conditions may not induce errors sufficiently great to account for the failure of the prediction. This is precisely the pattern of investigation revealed in Newton's 1672 paper. And Leverrier's initial work on Neptune follows the same pattern except for its use of rough calculations to replace actual empirical variation. In short, *it may never be necessary in the entire life history of a particular theory for the full apparatus of the theory to be invoked in a crucial test.* The whole process of falsification can be carried out using approximative forms.

IDEALIZATIONS

A number of important and basic scientific theories are approximative in a sense going far beyond that implied in the previous section. Duhem, for one, believed that mechanics was *essentially* approximative in such a way that the ideas of truth and falsity could not meaningfully be applied to its laws.[17] Duhem's point is a good one, but overstated. It *is* meaningful to speak of the truth and falsity of the laws of mechanics. But they are, as he suggested, approximative in a fundamental way. Let us try to say why.

Any theory which invokes ideal conditions among its scope assumptions is bound to be approximative in the strong sense. And Newtonian mechanics definitely has this characteristic. The Newtonian laws of motion—even Kepler's laws—all assume scope conditions which are not as a matter of fact present in nature. They do this *deliberately.* It is not a mere accident.

Newton's first law is a classic case: "In the absence of impressed forces, any body will move rectilinearly with an acceleration of zero." It is physically impossible, however, to produce the circumstance of there being no impressed forces acting on a body. Or, at least, if such a circumstance is produced we shall have no way of finding out what the acceleration of the body is since there will be no spatial

---

[17] Pierre Duhem, *The Aim and Structure of Physical Theory,* tr. by Philip Wiener (Princeton: Princeton University Press, 1954 (1914)), p. 168: ". . . the laws that a physical science, come to full maturity, states in the form of mathematical propositions . . . are always symbolic. Now a symbol is not, properly speaking, either true or false; it is, rather, something more or less well selected to stand for the reality it represents, and pictures that reality in a more or less precise, a more or less detailed manner. But applied to a symbol the words 'truth' and 'error' no longer have any meaning; so, the logician who is concerned about the strict meaning of words will have to answer anyone who asks whether physics is true or false, 'I do not understand your question.'" And again (*Ibid.,* p. 171.): ". . . every physical law is an approximate law. Consequently, it cannot be, for the strict logician, either true or false . . ."

reference points with respect to which its motion can be gauged. Attempts to simulate the condition of no forces (i.e., dynamic closure) by reducing the vector-sum of forces to zero do not quite do the trick. For the physical criterion for saying that the net forces are zero is that the body behaves inertially; i.e., that it remains at rest or remains in motion at constant velocity.

Kepler's laws of planetary motion are another case in point. "All planets move in elliptical orbits with the sun at one focus." But only if gravitational attractions of all other bodies in the universe are ignored! To produce the assumed conditions one would have to perform the impossible task of isolating the planet and sun as the only objects in the entire universe and then measure successively the relative positions of the two objects after the planet had been given an appropriate initial velocity. The mere description of the process involves inconsistency. For who shall perform the measurement? And with what? If no other objects are allowed in the universe it is difficult to see what can be meant by 'observation.' One is driven to invoke Laplace's Demon (or Napoleon's God, depending on one's theology) to carry out the observation and calculation.

Ernst Mach, perhaps more than any of his predecessors, felt the strain of this idealization and sought to reduce mechanics to a more solid basis. He succeeded in making hash of the theory. He began by construing the laws of motion as being compounds of empirical and definitional components.[18] The first law is taken as a definition of 'iner-

18 Ernst Mach, *The Science of Mechanics,* tr. by Thomas J. McCormack (LaSalle, Ill.: The Open Court Publishing Company, 1960), p. 266.

tial frame,' the second of 'force,' the third of 'mass.' These are then supplemented by appropriate empirical statements about the conditions under which masses are equal, etc. In the case of the first law, the empirical statement required to supplement the definition of inertial frame is the assumption that the coordinate frame located at the center of mass of all bodies in the universe (taken collectively) is an inertial frame.

Another way of stating the nature of Mach's program is to note that he is requiring so-called "operational" definitions for all of the concepts of the theory. This means that every boundary condition must ultimately be tied down to a specific measuring process which will decide whether and in what way the boundary condition is met. The criterion for a system's being an inertial frame is that no accelerations occur when a test body is placed by itself in the frame of reference. Whether this condition holds empirically or not for the frame designated by Mach must be determined by measuring the actual motions of a test body with respect to the center of mass of the universe. Thus, the question whether something is an inertial frame or not is effectively reduced to the problem of making measurements with an accelerometer—or, more simply, to a combination of time and distance measurements.

The thrust of Mach's approach is not unlike that of the H–D theorist. His arguments against the first law of motion as an "untestable" and metaphysical principle lead him along the same route we have just traveled in trying to adapt the H–D account to approximative inference. The fact that the conditions envisioned in the law (viz., dynamical closure of a one-body system) cannot be realized in nature, but only approximated, leads him to ask for a precise way of testing whether the law is true. This, in turn,

carries him on to an attempt to define the idea of dynamical closure operationally; and before he is finished we are
led to the conclusion that the only dynamically closed system is the totality of bodies in the universe. For this alone
constitutes a genuine inertial frame.

The practical absurdity of Mach's solution is self-evident.
A simple calculation of the trajectory of a hockey puck on
an ice rink requires that we refer the puck's motion to the
center of the earth, the earth's motion to the center of the
sun and the sun's motion to the universal center of mass.[19]

---

[19] Newton's *Principia, op. cit.,* p. 13: Commenting on the First
Law of Motion Newton says: "Projectiles continue in their
motion, so far as they are not retarded by the resistance of the
air, or impelled downwards by the force of gravity. A top,
whose parts by their cohesion are continually drawn aside
from rectilinear motions, does not cease its rotation, otherwise
than as it is retarded by the air. The greater bodies of the
planets and comets, meeting with less resistance in freer spaces,
preserve their motions both progressive and circular for a much
longer time." And N. R. Hanson, *Patterns of Discovery, op. cit.,*
p. 96: ". . . the empirical grounds for asserting the first law
are events like slipping on polished floors, or observing how
a round rock moves across ice with but slightly diminishing
velocity until it slows to a halt. When the first law statement
seems not to hold, the reason can always be found: ground
glass on the ice, perhaps, or the discovery that the rock is a
lodestone, etc. The law encapsulates and extrapolates much
information about events, yet it seems beyond disconfirmation:
it could not but be true." And, finally, R. B. Lindsay and
Henry Margenau, *Foundations of Physics* (New York: Dover
Publications, 1936, 1957), pp. 88–89: "Many elementary textbook writers content themselves with observing that when a
hockey puck slides on ice, the smoother the ice the farther the

In other words, every mechanical calculation involves implicit reference to every body in the universe! The simple expedient of regarding the puck's motion as an approximation to inertial motion is eliminated by Mach's argument against the "Newtonian" form of the first law of motion: a true closed system exists nowhere in the universe (excepting the system consisting of all bodies in the universe), hence it cannot be claimed that the puck is a true closed system dynamically. Hence—on the Machian view—the law of inertia cannot be applied directly to the case. In practice, of course, Mach relented and allowed that the center of gravity of the solar system could be taken as representing a probable inertial reference point. But even this concession does not get us very much closer to the hockey puck.

My intention here is not to construct a general critique

———————————————————————————————

puck travels with a given blow before coming to rest. They then ask us to imagine that the ice becomes in the limit perfectly smooth—an ideal surface which has no effect on the puck. The assertion is then made that the puck would continue indefinitely in a straight line with constant velocity. As a suggestive illustration one can hardly criticize this, though it is well to point out that the surface must be idealized beyond the limit suggested, i.e., it must be made infinite in extent and, more important still, must be flat, i.e., cannot be on the surface of the earth. . . . In other words, the illustration which sounds at first not bad proves very unfortunate on closer inspection. Probably much the same thing would be true of any attempted large-scale phenomenal illustration of the first law of Newton. It is questionable procedure to try to clear up the meaning of a fundamental physical law by giving an illustration which breaks down badly on the slightest questioning."

of the Machian reinterpretation of classical mechanics.[20] Instead, I want to call attention to a very fundamental point about classical mechanics which Mach's reformulation throws clearly into relief. The point is that the idealized notion of "dynamically closed system"—of a system subject to the action of no external forces—is an irreducible element of the theory. Any attempt to define the notion operationally leads one in the end to bring in all bodies in the universe since, by Newton's law of gravitation, any body left outside our alleged closed system will act on the system with a force inversely proportional to the square of the distance from the outside body to the center of gravity of the system. So they have all got to be included.

The trouble with including all bodies in this way is that it renders the theory physically meaningless, i.e., unfalsifiable.[21] For suppose that a test body is shown to accelerate markedly with respect to the center of mass of the universe even though under the gravitational or magnetic influence of no nearby objects. Will this lead to the overthrow of classical mechanics? Not in the Machian interpretation. There the occurrence simply shows that the empirical statement associated with the first law is false: the frame of reference situated at the center of mass of the universe is not an inertial frame. But according to the second law—construed as a definition of 'force'—the presence of the acceleration also shows that there is a force of a certain magnitude acting on the body! We must therefore swallow

---

[20] For such a critique see Ernest Nagel, *The Structure of Science* (New York: Harcourt, Brace and World, 1961), Chapter 7.

[21] Refer again to the quotations given in Footnote 19 above for an illustration of how this occurs.

hard and accept the result that a perturbing force (let us call it an "X-force" to suggest the appropriate air of strangeness) acts in such a way as to produce the observed anomaly and to render the center-of-mass reference frame non-inertial. Only the first law is falsified. The rest of the theoretical edifice—plus the mysterious X-forces—remains. And, of course, with X-forces one can explain everything from the advance in the perihelion of Mercury to terrestrial levitation.

We are skirting here on the fringe of Newton's old problem about absolute space. Mach identifies Newton's imaginary absolute reference frame with a physical frame and reinstates all of Newton's difficulties in slightly different language. Either way, the result is the same: the theory of mechanics loses all physical content and becomes an ethereal body of speculations on a world that never was nor ever will be.

The plain historical fact, however, is that Newtonian mechanics was a falsifiable, empirically meaningful theory! Its rejection in the 20th century stems not from revulsion at its conceptual untidiness but from empirical counterevidence. From the Machian standpoint, from the H–D theory standpoint, from Newton's own absolute-space standpoint this is unintelligible. Some other explanation must be given.

Recall Duhem's thesis that the laws of physics are all approximations. What bearing does this have on a proper understanding of Newtonian mechanics? Just this: *Newtonian mechanics is a falsifiable, empirically meaningful theory because it is essentially approximative.* All explanations and predictions based on the theory are approximative inferences in which the assumption of dynamic closure for the system under consideration lurks in the background. No actual inference ever begins with a complete list of initial conditions for all bodies in the universe. Some forces and gravitational attractions are always assumed to be "insignificant" or "too minute to be considered."

What happens in a case like the advance of the perihelion of Mercury is that the approximative inference turns out to be completely justifiable even though it generates faulty predictions. All *relevant* small forces can be taken into account *via* rough calculations and they are not sufficient to explain the anomaly. There is no need to drag Arcturus into the picture here, since its influence on Mercury will be demonstrably smaller than that of the earth's moon. And if Arcturus need not be considered, why go further? Mach's ultimate inertial frame and Newton's absolute spatial envelope are simply irrelevant. We do not need to know where the center of mass of the universe or the imaginary "absolute center" is in order to falsify the theory.

Newtonian mechanics is not the only theory within which idealized scope conditions render all inferences approximative. Classical thermodynamics—to the extent that it purports to deal with "perfectly" insulated thermal systems—is another such theory. And we can add gas dynamics with its "ideal gases," hydrodynamics with "ideal fluids" and elementary statics with its "ideal beam balance" to the list. The logic of falsification of these theories is not the matter of "elementary logic" mentioned by Braithwaite. It is not simply a question of deducing a false prediction and then waiting for a plausible hypothesis. The deduction of the false prediction *via* approximative inference is followed by a searching investigation to determine whether the inference was justifiable. This involves a canvass of all relevant sources of error; which means, in turn, that the theory must be retained until the investigation is complete.

The use of idealized systems and idealized boundary conditions in physical theories is intimately bound up with a familiar logical puzzle which has exercised philosophers for some time; namely, the problem of specifying truth conditions for counter-to-fact conditional statements. This problem arises in idealized theories for the simple reason that *any law describing the behavior of an ideal system has the logical force of a counterfactual conditional.*

This patently obvious fact seems to have been largely overlooked by philosophers concerned with the explication of counterfactuals. By concentrating on the counterfactuals implied or supported by physical law statements ("If this were copper then it would conduct electricity") they have overlooked the intrinsic counterfactuality of a whole class of physical laws. True, Newton's first law of motion has been singled out for attention on the basis of its counter-to-fact character. But it has been tacitly assumed that the first law is an exception. In point of fact, it is more like the norm of a physical law. The grammatical form of any solution of a dynamical problem is: "In a system of n bodies in which no external forces are assumed to be acting . . ." —which can equally well be read: "If there *were* a dynamically closed system of n bodies, then the initial conditions $a, b, \ldots, c$ would determine succeeding states of the system in such and such ways." Similarly: "If there *were* an ideal gas, its pressure, volume and temperature *would* be related as $PV = T$." And: "If there *were* an ideal beam balance, it *would* achieve equilibrium only in the case where $W_1S_1 = W_2S_2$, where the W_i are weights suspended at distances S_i from the fulcrum."

All of these assertions are at least implicitly counterfactual because of the kind of theory in which they are imbedded. In virtue of other propositions in the overall theory we know that the antecedents of these laws cannot be fulfilled. Newton's law of gravitation plays this role with respect to dynamical closure; and principles concerning friction and molecular behavior of gases likewise show the physical impossibility of ideal gases and beam balances. We know—*a priori* as it were—that the boundary conditions cannot be fulfilled exactly in any empirical case.

Some philosophers have come perilously close to concluding that any counterfactual conditional must be a metaphysical utterance; i.e., that no such conditional statements are falsifiable on the basis of empirical evidence. Perhaps this is the kind of thing Duhem intended when he claimed that physical laws *qua* approximations cannot be either true or false. But this is simply not so. A counterfactual conditional is often an empirically meaningful statement—only one whose truth conditions are not immediately specified by the meanings of the constituent terms. As a prototype, consider the counterfactual "If this were July 4th (when in fact it is only July 1st), then today would be Tuesday." This statement is clearly falsifiable. All one need do is look at the calendar and count. If today is July 1st and a Monday then the counterfactual is false. In place of it we would have to substitute: "If today were July 4th then it would be *Thursday*" or "If today were July *2nd* then it would be Tuesday." Note that the feature which enables us to assess the truth or falsity of these subjunctive conditionals is the existence of a rule or law which links up the antecedent and consequent in a certain way. The rule allows us to say precisely how an alteration in the main antecedent term ('July 4th') will affect the main consequent term ('Tuesday'). The relationship is a simple function linking calendric numerical dates with days of the week.

The situation with idealized physical theories, of course, is a good deal more complicated than our simple example. But the principle is basically the same. The existence of some kind of functional relation between boundary conditions and the effects occurring in the system guarantees that we can falsify the subjunctive conditional law. In fact we can say quite generally that in any counter-to-fact conditional where a causal or functional link can be found between antecedent and consequent the truth conditions of the conditional will be rather crisply definable. The one qualification that must be added is that in an essentially approximative theory (like classical mechanics) it is often impossible to eliminate all vagueness whatever without rendering the theory physically meaningless. For, as we have seen, the attempt to eliminate boundary conditions in favor of initial conditions of a wider system of objects leads in the end to inclusion of the entire universe in one monstrous dynamical unity. In practice, the elimination of "outside influences" proceeds experimentally and *via* rough calculations, not explicit inclusion of all conceivable influences in a superhuman deductive inference.

One final matter deserves our attention here before we leave the subject of idealizations. That is the queer feature of idealized theories which makes them stand or fall as a whole rather than piecemeal. Why is it that the demise of classical mechanics gives rise to a wholesale conceptual revision in physics? Or, to put it another way, why does Newton's discovery of the elongation of the spectrum not lead to the complete overhaul of geometrical optics?

The answer is to be found in the counterfactuality of the idealized theory's laws, not—as the H–D theorists claim —in the deductive connection between hypotheses (laws) and observation statements. When an anomalous observation is found to be inexplicable using the laws of an idealized theory, it remains an undecidable question which of the laws is "at fault." This is not because we have a deductive inference going one way from hypotheses involving theoretical terms to observation statements involving only empirical terms. In an H–D system of that sort it might still be possible to "assign the blame" on other grounds. *But in an idealized theory there is absolutely no way to fix the blame.*

Ask this question: does the advance in the perihelion of Mercury prove that were Mercury under the influence of no external forces it would accelerate nevertheless? In other words, is the law of inertia falsified by the advance in the perihelion of Mercury? The question just cannot be decided. *The rule which we used to infer from ideal conditions to empirically real ones and back again has been called into question by the anomaly.* It is as though we were suddenly apprised of the possibility of there being two Tuesdays in a given week and then asked to decide whether if today were July 4th it would be a Tuesday. Without a whole new conceptual framework (calendar) accommodating this business about two Tuesdays we cannot answer the question.

The "rule" which breaks down in classical mechanics is the law of universal gravitation. We have as much (and as little) right to say that the law of inertia is falsified by Mercury's behavior as to say that the law of gravitation is. Or is it Kepler's laws? ("Mercury wouldn't really move in an elliptical orbit even if it were, with the sun, the only

other body in the universe.") We are at a loss to say. For
none of the counterfactuals asserting the falsity of the laws
of planetary motion has determinate truth conditions unless the law of gravitation and second law of motion are
assumed. How, after all, could one go about estimating
the behavior of Mercury in an isolated, one- or two-body
system without "subtracting" the gravitational attractions
of the nearby massy bodies? The very calculations presume
the validity of the law of gravitation (or its dynamical
equivalent in the Hamilton formulation of mechanics).

It is a gross oversimplification to describe this situation
as a case in which a set of hypotheses has been falsified
as a whole and we are unable to discern which one of them
is false because some include "theoretical terms." Granted
that the law of gravitation and the law of inertia and
Kepler's law are all involved in deduction of predictions
about the behavior of Mercury. It is still conceivable that
we might be able to determine empirically which law is
at fault if only we could approximate the boundary conditions for each law in turn. The trouble is that 'approximate' no longer has a meaning! For unless we can use the
law of gravitation as a calculating tool we cannot say of
any empirical setup that it is a good approximation to a
closed system or a bad one. Here, then, is the source of
the theory's essential unity and the cause of its being falsifiable only in total, not piecemeal. The presence or absence
of "theoretical terms" in the laws has nothing to do with it.
This point ought to have been clear right from the start.
Wasn't Newton able to decide empirically which law was
at fault—Snell's law or the rectilinear propagation principle? And doesn't geometrical optics involve "theoretical
terms" to just as great a degree as Newtonian mechanics?
The real difference is that the boundary conditions of the

laws of geometrical optics *are* satisfiable in nature. Again: "theoretical terms" have nothing to do with it.

ANOMALIES AND THE STRUCTURE OF THEORIES

Let us now bring this discussion of the nature of theories full circle back to the idea of anomaly. For the kernel of what has been said in the last few pages can be brought to light most forcefully by relating it to what was said earlier about the "definition" of anomalies.

The process which we earlier described as the definition of an anomaly is none other than the process which we have lately been calling 'justification of an approximative inference.' Because physicists customarily draw inferences on the basis of less than complete evidence anomalies nearly always present themselves in such a form that the presence of a contradiction between theory and observation-descriptions cannot be immediately ascertained. Only after the possible sources of error allowed by the theory have been fully explored, only when all explanatory resources of the theory have been brought to bear, can we conclude once and for all that theory and observation are irreconcilable. Thus, for instance, the researches of Newton and Leverrier in trying to pin down the anomalies they later succeeded in explaining involved the very techniques we have attributed to justification of approximative inference: rough calculations to determine the limits of error introduced by "contingencies," as Newton called them; experimental variation of conditions to the same end; and comparison of multiple observations (Leverrier) as a means of ruling out these same "contingencies." In both cases we have, as the starting point of the investigation, an approximative inference leading to grossly erroneous predictions.

In both cases, the discovery of an explanation provides us with a better approximation for future use.

It goes without saying that the whole idea of "defining the anomaly" is meaningless in the H–D framework. Look again at Braithwaite's description of the situation in which an anomaly arises: "What experience can tell us is that there is something wrong somewhere in the system; but *we can make our choice* as to which part of the system we consider to be at fault." [22] The methodological implications of this are startling! A physicist who follows the Braithwaite prescription when faced with an apparent counterinstance ought to sit back, light his pipe and take his pick as to the faulty hypothesis. In fact, he is far more likely to go scurrying off to the laboratory or observatory for further data. The only time the Braithwaite prescription holds good is when all of the data are in and we are dealing with an idealized physical theory like classical mechanics.

The H–D account thus omits mention of all of the scientific activity between (a) the deduction of a faulty prediction and (b) the certification that the theory has been refuted. More important than this mere genetic oversight, however, is the omission of any logical account of how a theory can be *corrected* via empirical investigation. Some of the most brilliant research in the entire history of science has been devoted to such problems, Newton's optical work included. The best account of this the H–D model gives us is that the laws of a corrigible theory must be low-level generalizations of observations and not really theoretical claims (viz., claims employing theoretical terms). It is diffi-

---

[22] Braithwaite, *op. cit.,* p. 19. My italics.

cult to see, however, that the principle of rectilinear propagation of light is any less theoretical than Kepler's laws or the law of universal gravitation. What are the special theoretical terms that set Kepler's laws and the law of gravitation off? Aren't all of the terms of these laws entirely comparable to the terms of the principle of rectilinear propagation?

Because it glosses over definition of anomalies and the use of approximative inference the H–D account of scientific theories also leads us to miss another important feature of physical theories. *To a great extent, physical theories preserve a record of their explanatory successes in their logical form.*[23] The approximative forms of various laws often represent earlier stages in the growth of a theory. Thus, for example, Kepler's own statement of the laws bearing his name does not take into account perturbations of the elliptical path by external gravitational forces. When Newton incorporates Kepler's laws into a wider dynamical theory (ideal) boundary conditions are added. Yet the laws continue to be used in their approximative form for various elementary calculations. The observations which are anomalous with respect to Kepler's own formulation are explained *via* the introduction of Newtonian and post-Newtonian correction factors.

The attempt to describe all of this in terms of a purely axiomatic system deprives the theory of any semblance of explanatory power. Physical explanation does not simply consist of deducing the appropriate observation statements

---

[23] This is one important source of the strong tendency in philosophy of science and history of science toward genetic fallacy and the confusion of the order of discovery with the order of justification.

from the appropriate laws—though that is certainly involved. The main thing is that the anomalousness of the observations be resolved. And this requires that we at least be able to say in what respect the observations were anomalous in the first place. *If Newtonian mechanics were really the kind of theory the H–D model represents we could not do this. Intratheoretical explanation would be impossible to achieve for the most part.* In a purely axiomatic theory we could never derive both a true and a false prediction about the same subject matter. In at least one of the forms of intratheoretical explanation, however, it is possible to do just that. The trick is that the inference to the false prediction is approximative while the inference to the correct prediction is either exact or, at least, a better approximation. Thus the *appearance* of inconsistency required for explanation can always be simulated without the theory actually being inconsistent.[24]

---

[24] As an illustration of this apparent inconsistency look at this remark by Laplace: "The moon moves in an elliptical orbit, of which the centre of the earth occupies one of the foci. . . . [Next paragraph, same page] The laws of the elliptic motion are very far from representing the observation of the moon; it is subject to a great number of inequalities, which have an evident connection with the position of the sun." Laplace's *System of the World,* tr. by Henry H. Harte (Dublin: At the University Press, 1830), Vol. 1, p. 31. Compare this with Duhem's famous remark (Duhem, *op. cit.,* p. 193): "Is this principle of universal gravitation merely a generalization of the two statements provided by Kepler's laws and their extension to the motion of satellites? Can induction derive it from these two statements? Not at all. In fact, not only is it more general than these two statements and unlike them, but it contradicts them."

THE H–D MODEL REVISED

The criticism we have brought against the H–D account must now be summarized and brought into focus in a revised view of the nature of physical theories. Since some of these criticisms concern the *form* in which theories are customarily expressed or the *procedures* of inference customarily followed in physical explanation and prediction it will be necessary for us here to separate carefully the pragmatic from the logical. In other words, it must be asked whether our criticisms show that the H–D account is defective as a logical model or merely that it fails to give a satisfactory *genetic* account of the growth of scientific knowledge and the practices of scientists. From the standpoint of the orthodox H–D theorist only the former kind of claim is of major significance. He does not conceive himself to be in the business of describing the *process* of scientific inquiry, only the *product*. (A perceptive reader will have noticed, however, that in the previously quoted remarks of Professor Braithwaite this tidy distinction is not maintained with all possible rigor. *Vide* his description of the *procedure* followed by scientists when predictions are falsified.)

What I wish to maintain here is that the H–D model *is* logically defective and not merely misleading methodologically and historically. At the same time, I wish to stress that it is entirely serviceable as a logical model of theories whose approximative character is only expository, not essential. The H–D account founders only on theories involving essential idealizations. *Either we give up the H–D model or we must concede that such theories are empirically unfalsifiable.*

Consider, for example, the pair of inferences drawn by

Leverrier from the basic laws of Newtonian celestial mechanics and leading to predictions of the future positions of Uranus. One inference is drawn *before* the discovery of Neptune, the other afterward. The first does not invoke initial conditions for Neptune and assumes that the system Sun-Jupiter-Uranus is closed dynamically. The second does include mention of Neptune and assumes closure for the wider system. Both times the closure assumption is known to be counter-to-fact. The predictions generated in each case are incompatible with one another. If we symbolize the respective closure assumptions as C_1, C_2, the appropriate data statements as D_1, D_2, and the basic mechanical theory as T, the inferences have the following appearance:

Inference One: $T\&C_1$ & $D_1 \rightarrow P_1$.
Inference Two: $T\&C_2$ & $D_2 \rightarrow P_2$.

P_2 is a set of predictions logically incompatible with P_1 but empirically accurate within limits of observational error. The crucial defect in the H–D account is that it provides no way of evaluating the relative soundness of these inferences. There is no way to judge which is the logically "correct" conclusion to draw from theory T about the behavior of Uranus, P_1 or P_2.

If the H–D theorist claims that Inference One is defective in including the false statement C_1, he is obliged also to condemn Inference Two on the same grounds since C_2 is known to be false as well. Only by recasting T in a Machian form, as we have seen, will it be possible to eliminate the false closure assumption—and that means transforming T into an unfalsifiable theory.

If the H–D account is really an adequate schematization of the logic of Inferences One and Two it would always be open to a physicist to excuse faulty predictions by simply saying "Oh well, that just shows what we already know: the system under consideration is not dynamically closed." A false prediction implies that one premise at least is false. But since C_1 and C_2 are known to be false to begin with we gain no new information from finding that P_1 is false.

This horrendous misinterpretation of the complex logical relationship between theory and observation statements must surely be rejected. Some more satisfactory description must be given of the sense in which Inference Two is preferable to Inference One. And that means specifying how the falsity of prediction P_1 provides information about which premises—T or C_1 or D_1—need to be re-examined.

The concepts of approximative forms of law statements and approximative inference provide a way of clarifying the situation while retaining what is obviously valid in the H–D account. We can put the matter this way: the approximative forms of a law can be put roughly into an ordering depending on which scope conditions (and how many of them) are suppressed. Or, what comes to the same thing, in classical celestial mechanics and similarly idealized theories closure assumptions can be ranked according to their degree of approximation (C_1 being, for instance, a lower order approximation than C_2). Such language is even commonly used in physics, where one speaks, say, of a two-body orbital solution as a first-order approximation and a three-body solution as second-order.

Inference Two, then has logical priority over Inference One in the sense that it represents a better approximative inference. It utilizes a closure assumption of higher order than Inference One or, to put it the other way round, it more nearly approaches the (essentially unrealizable) completeness of scope conditions required by the Newtonian laws of motion.

In order to reformulate the H–D model in such a way as to incorporate these ideas it is necessary to drop the assumption that the fundamental principles and laws of a theory should be expressible as ordinary statements. Instead, we shall assume that any principle may be construed as a set of statements—usually laid out in roughly the order of their degree of approximateness—in which more and more of the scope conditions are suppressed. Snell's law, Kepler's laws, the principle of rectilinear propagation and the like are examples to bear in mind.

A theory, then, is a set of such explanatory principles (statement sets) together with the "families" of explanations and anomaly contexts in which they occur. The logical relations are depicted in the following schematic diagram. The P_i represent basic principles of some theory or other: say, geometrical optics or Newtonian mechanics. The E_j are explanantia and the A_k anomaly contexts in which the principles play a role. Although only one anomaly context is paired with each explanation in the schema it is obvious that there may be many in an actual theory.

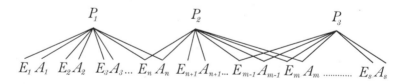

The significance of this more complex account of the structure of theories is that it captures not only the logical features of idealized theories glossed over in the H–D account but also serves more adequately to characterize the actual practice of scientists in handling both types of physical theory. Basic principles are allowed a variety of formulations depending upon the degree of "looseness" required for practical purposes. For rough calculations a rough formulation will do. When accuracy and precision are needed, a tighter statement involving more of the scope or boundary conditions is employed. We emphasize again, however, that in the case of idealized theories this is not merely an expository or pragmatic device. It is the only way that the appropriate inferences can be drawn.

In an idealized theory, then, looseness of formulation is essential rather than accidental. *Only if the principles* P_i *are stated more or less loosely can they appear in all of the explanatory arguments they actually appear in.* A "precise" statement of Kepler's laws, for instance—one spelling out the exact scope conditions under which the laws apply without error—could not fit into *both* an approximative inference about the planet Uranus and a separate exact inference in which corrections are introduced.

Of course, if we are willing to regard a theory *merely* as a predictive or calculation scheme none of this makes a particle of difference. Predictive inferences do not require anomaly contexts at all. Only explanations do. It is difficult to believe, however, that scientific theories are constructed only for purposes of prediction. They are designed, rather, to solve problems about empirical subject matter—to resolve anomalies and provide explanations. Their genetic growth, as we have said, comes usually through a process of accumulation of new explanations involving the (loosely formulated) general principles. This history of growth is written indelibly into their logical form.

4

The A Priori in
Statistical Theories

The leitmotif of modern science is the use of statistical
laws and probability concepts. Not only in physics—where
quantum mechanics, thermodynamics and statistical mechanics come readily to mind—but also in the biological
sciences, probabilistic explanations and reasoning have become the rule rather than the exception.[1]

The growing use of statistical laws—as opposed to general or causal laws—raises (or perhaps one should say
"re-raises") certain important philosophical issues about
the meaning and interpretation of the concept of probability. Especially in quantum mechanics clarification is

---

[1] There is no intention here of suggesting—as Jacob Bronowski
does in *The Common Sense of Science* (New York: Vintage
Books [No date]), p. 95 *et passim*—that chance constitutes a
new metaphysical principle in modern thought. Much of what
Bronowski refers to as "chance" comes under the heading of
what philosophers have always called "apparent chance," i.e.,
chance arising from limits on human information and knowledge rather than from the nature of Nature.

needed about the ideas of chance, frequency, likelihood, and so forth.

The aim of this chapter is to consider some of the interpretations of probability currently of interest to philosophers of science and to trace out some of the implications of these theories for our understanding of the foundations of elementary modern physics. After considering three such interpretations—the classical, orthodox frequency and logical—we shall try to propose an alternative reading which squares more fully with the actual practice of contemporary science and which may also serve to give a more unified account of the nature of statistical theories in physics, biology and elsewhere. Objections to this alternative will be considered and we shall conclude by defining the notion of statistical anomalousness.

THE INTERPRETATION OF PROBABILITY

The classical interpretation of probability—the one familiar to dice players, mathematicians and others of dubious character—was first formulated by French philosophers and mathematicians in the 17th and 18th centuries.[2] The culmination of their work is to be seen in the Marquis de

---

[2] A thorough discussion of the history of the (non-mathematical) foundations of probability occurs in J. T. Merz, *A History of European Thought in the 19th Century*, four volumes (New York: Dover Publications, 1965 (1904–1912)), Vol. II, Chapter XII. See also A. Wolf, *History of Science, Technology, and Philosophy in the 18th Century*, two volumes (New York: Harper, 1961 (1938)), Vol. I, Chapter I.

Laplace's justly famous *Philosophical Essay on Probabilities* (1814).[3] As Laplace sees it, the central problem in defining probability is to determine some way of assigning basic probability numbers—real numbers lying between zero and one—to certain occurrences which we may refer to as "simple events."[4] Once the probabilities of these simple events are fixed, it will be possible—using the rules of the mathematical theory of probability—to calculate the probabilities of compound events. Thus, if it is possible to assign probabilities to the simple events of getting head or tail on the flip of a coin, the probabilities of compound events like strings of six straight heads can be computed directly Basically the problem that Laplace set himself to solve (viz., the problem of interpreting the idea of probability for some set of basic events) is the same problem philosophers still grapple with today.

Laplace chose to use as a foundation for his interpretation a rule or principle which has had a long and illustrious history in rationalistic philosophy: the principle of suffi-

---

[3] Pierre Simon, Marquis de Laplace, *A Philosophical Essay on Probabilities,* tr. from the sixth French edition by F. W. Truscott and F. L. Emory (New York: Dover Publications, 1951). The mathematical details of the theory of probability are suppressed in this foundational work but can be found in Laplace's *Théorie analytique des probabilities.*

[4] *Ibid.,* p. 6: "The theory of chance consists in reducing all the events of the same kind to a certain number of cases equally possible, that is to say, to such as we may be equally undecided about in regard to their existence, and in determining the number of cases favorable to the event whose probability is sought."

cient (or, if you prefer, insufficient) reason.[5] According to this principle, if we know that there are n possible outcomes to the test or experiment we are going to perform and if nothing is known which favors any one possibility over the others, then we assign the probability $1/n$ to each of them. Thus, since we have insufficient reason to expect heads over tails on the flip of a coin we make the probability of each event $\frac{1}{2}$. (This, of course, is based on the assumption that a coin landing on edge will not be regarded as a bona fide toss.)

Laplace's method of assigning probabilities is a form of subjective probability theory. By this we mean that it assumes implicitly a certain level of ignorance on our part concerning the causes of the various outcomes and measures our "reasonable degree of belief" in that situation of ignorance. Laplace himself was convinced that if only we knew enough about relevant conditions it should be possible to predict with certainty whether heads or tails would

---

[5] The most famous proponent of the principle, of course, is Leibniz. For a recent and lucid discussion see Nicholas Rescher, *The Philosophy of Leibniz* (Englewood Cliffs, N. J.: Prentice-Hall, 1967), p. 25ff. The full statement of the principle asserts that any statement is true if and only if it is analytically true; hence, for the description of any event there must exist axiomatic foundations from which a description of the event can be deduced. In the case of probabilistic reasoning, our inference proceeds from less than "sufficient reason" which is to say that a prediction of the event cannot be deduced with certainty from analytically true premises. When "sufficient reason" is symmetrical with respect to the occurrence or non-occurrence of the event the probabilities of occurrence and non-occurrence must be taken as equal.

turn up on the next toss; and Newtonian mechanics certainly gave good grounds for believing this.[6] Because we
do not know about the initial and boundary conditions
affecting the coin—the force and direction of the toss, the
wind currents encountered, etc.,—we are forced to fall
back on probabilities. But the element of chance involved
is not, for Laplace, an objective part of Nature itself.

The principle of sufficient reason is a most beguiling
piece of metaphysics. By magic, as it were, it proposes to
turn our ignorance into knowledge. Like most magic, however, its basis is largely that of illusion. For the fact is that
when applied without restriction the principle leads us
into contradiction. Consider, for instance, the problem of
assigning probabilities to *pairs* of tosses of coins. Proceed-
ing in the usual Laplacian way we will find that the probability of getting two heads calculates out to be $1/4$, two tails
likewise, and the probability of one head and one tail—
since it can occur two ways—will have a probability of $1/2$.
All this on the assumption that heads and tails are equally
likely.

Suppose, however, that we begin with three "simple"

---

[6] Laplace, *Philosophical Essay on Probabilities, op. cit.,* p. 4:
"Given for one instant an intelligence which could comprehend all the forces by which nature is animated and the
respective situation of the beings who compose it—an intelligence sufficiently vast to submit these data to analysis—it
would embrace in the same formula the movements of the
greatest bodies of the universe and those of the lightest atom;
for it, nothing would be uncertain and the future, as the past,
would be present to its eyes. The human mind offers, in the
perfection which it has been able to give to astronomy [New-
tonian celestial mechanics], a feeble idea of this intelligence."

events: (1) two heads, (2) two tails, (3) one of each.[7] If these
are construed as the simple events, application of the principle of sufficient reason tells us that each of the three has
a probability of 1/3. This is manifestly incompatible with
the usual assignments.

We know, of course, that the second way of assigning the
probabilities is—in some sense—incorrect. Defenders of
the classical interpretation would say that we had not really
chosen the *simple* events. (This immediately leads one to
wonder how we can really ever be sure that we have chosen
the simple events. After all could it not be that there is
something simpler than getting heads or tails?) Critics of
the classical interpretation, however, take a different tack.
Among some of them, at least, this kind of paradox signals
the need for a different fundamental interpretation—one
which does not rely on the principle of sufficient reason
at all.[8] Such an interpretation is the *relative frequency* theory of probability.

According to the frequentists, the trouble with the second way of assigning probability numbers to tosses of coins
is simply that *as a matter of fact* lengthy experiments with
pairs of tosses will lead us to the conclusion that the head-
tail combination shows up more nearly half the time than
one-third of the time. And we should also find, as a matter
of fact, that in the long run double-heads comes out about

---

[7] See Arthur Pap, *An Introduction to the Philosophy of Science*
(New York: Free Press, 1962), pp. 191–92.

[8] This is not, to be sure, the only criticism of the classical
theory which is employed. See Pap, *ibid.*, and John Maynard
Keynes, *A Treatise on Probability* (New York: Harper and
Row, 1962 (1921)), Chapter IV, for a sampling.

¼—not ⅓—of the time. The frequentist, in other words, believes that our real grounds for assigning probability values are the actual experiences we have with objects and these experiences will lead us to rule out the second method of assigning probabilities. Basically, the frequentist *defines* probability as the ratio (or the limit of the ratio) of favorable cases to total cases in the long run.[9] The basic assignments of probabilities are therefore matters of

---

[9] The most precise mathematical formulation of the idea of relative frequency is that of Richard von Mises. Von Mises employs the idea of a "collective" which is defined as follows: "A collective is a mass phenomenon or an unlimited sequence of observations fulfilling the following two conditions: (i) the relative frequencies of particular attributes within the collective tend to fixed limits; (ii) these fixed limits are not affected by any place selection." Such a collective would be, for instance, the usual string of heads and tails (HHTHTTTHH-THTTTTHHTHHTT . . .) which one might come up with in tossing a well-made coin. Here—it can plausibly be assumed —the relative frequency of heads tends toward some fixed limit as the sequence extends in length and condition (ii) also seems to be satisfied (i.e., if we select arbitrary sub-classes from the list of tosses the relative frequency of heads in the sub-classes also tends toward the same limit as the length of the sequence becomes infinite). The limit value toward which all partial relative frequencies—relative frequencies in the sub-classes—converge is defined to be "The probability of heads in the collective." See Von Mises' *Probability, Statistics and Truth* (New York: The Macmillan Co., 1957), p. 28. For a critical discussion see Ernest Nagel, *Principles of the Theory of Probability*, Vol. I, No. 6 of *The International Encyclopedia of Unified Science* (Chicago: The University of Chicago Press, 1939), p. 51ff.

fact for the frequentist, rather than matters of pure reasoning. We shall pass over here the subtleties and difficulties connected with the frequentist's idea of "the long run" and his use of the idea of a limit of a ratio or sequence of ratios.[10]

Two points regarding the "orthodox" relative frequency interpretation must be borne in mind. In the first place, orthodox frequency theory requires that all probabilistic statements refer to *classes* of events or things, rather than individual cases. An individual case has no such thing as a frequency and hence no such thing as a probability.

The second important fact about probabilities of a frequency sort is that they are constantly subject to revision as new experience piles up. Where the classical Laplacian probability assignments are to be conceived as invulnerable to anything that happens in nature, the numbers assigned to basic events by the frequentist must be altered each time such a basic event *e* occurs or fails to occur. (We assume, of course, that frequencies for an entire set of mutually exclusive and collectively exhaustive events are given and that *e* is one of the events.) Even if it is assumed that the frequencies observed will in the limit converge to a single value, our *estimates* of that limiting value must always be modified by the facts as they come in. If a coin shows heads 1,000 times and tails 2,000 the classical Laplacian is free to assert that the probability of heads is $\frac{1}{2}$. The frequentist, however, is obliged to estimate the probability at $\frac{1}{3}$ pending further experience with the coin.

The chief advantage of the frequency interpretation of probability is that it opens the way for all kinds of applica-

---

[10] *Ibid.*

tions of probability which the classical theory in and of itself cannot accommodate. Insurance statistics are a good example. What events, after all, could we regard as the simple events in seeking to calculate the probability that a 60 year old man will die before age 70? The frequency with which such men die automatically gives us a measure of the probability. Nevertheless, the frequency interpretation of probability brings with it certain difficulties almost as serious as those which plague the classical interpretation of Laplace. We shall look at some of these in a moment.

THE LOGICAL INTERPRETATION

Rudolf Carnap, the patriarch of American analytic philosophers, wrote in a well-known paper[11] a few years ago that when one asks scientists whether a single idea of probability is used in science one always gets a yes answer. But when one asks for a definition, Carnap went on to say, two distinct meanings are offered. One of these, of course, is the idea of relative frequency. The other, however, is a concept more closely aligned with the classical viewpoint. After analyzing the matter carefully, Carnap came to the conclusion that both notions were really essential in scientific practice and that the classical interpretation in some form had to be retained. He proposed, however, to give it a radical renovation. The refurbished classical interpretation he referred to as the Inductive or Logical Theory of probability. Others have called it the A Priori Theory and Carnap

---

[11] Rudolf Carnap, "The Two Concepts of Probability," in Feigl and Brodbeck's *Readings in the Philosophy of Science, op. cit.,* pp. 438–55.

speaks of it often as the Degree of Confirmation concept. I shall try to stick with the terminology of "Logical Theory" but shall use the symbol 'C' to represent probability assignments made in this manner.

Briefly, the logical interpretation is as follows: statements of probability in dice games and the like—to the extent that they are regarded as invulnerable to empirical disconfirmation—should be assimilated to statements about the probability of hypotheses or theories on given evidence. Such statements, in other words, express a logical relation between the evidence e and the hypothesis or prediction h.[12] They are *a priori* and independent of experience simply because they are expressions of logical relations. If the relation is one of deductive connection—i.e., if the evidence we have deductively entails the hypothesis —then the probability will, of course, be 1. If evidence and hypothesis contradict one another, the probability is zero. Usually, however, the probability will be something intermediate between these extremes. The inference from e to h, in other words, will be inductive.[13]

Carnap's formulation does not really rely on the principle of sufficient reason, either. For he allows that we may

---

[12] Symbolically, they are written '$C(h, e) = x$' where x is some real number between 0 and 1 inclusive.

[13] By definition, an inductive argument is one in which the premises lend some degree of probability to the conclusion. These are normally distinguished from deductive arguments in which the truth of the premises is supposed to guarantee the truth of the conclusion. Later in this chapter I shall argue that the distinction is illusory and that inductive arguments are merely deductive ones in which one or more of the premises is an unproved hypothesis.

choose the probabilities of the "simple" events in any self-consistent way we like. There are, in a word, infinitely many inductive logics that we can set up—including the one where the probability of a pair of heads is $1/3$, rather than $1/4$.[14] We cannot mix these methods of weighing evidence, of course, for that would lead to contradiction. But it is a matter of convention which one we actually choose to employ. Just as there are many systems of geometry that a scientist might elect to use, there are also many systems of inductive or probabilistic logic for weighing the probability of hypotheses. All such logics or methods for assigning basic probability are equally valid a priori, though obviously the one we find most useful is that which agrees with the assumptions of Laplace. It is a bit like the use of different bases in arithmetic: all are equally valid, but base 10 enumeration seems most useful for most purposes.

How the basic assignments are made depends, in the logical interpretation, on symmetries in the structure of the descriptive language a scientist uses. The way in which the fundamental predicates of the language (e.g., 'is a head' or 'is a tail') are related, together with a conventional "weighting" of these predicates as regards their probability of occurrence, uniquely determines the probability of all hypotheses about what will happen, say, on the next toss or the next five tosses of a coin. In a sense, then, Carnapian logical probability substitutes the symmetry of linguistic descriptions for the *de facto* symmetry of physical coins and dice which the Laplacian appealed to. This is a remarkable and valuable example of what is often called the "linguistic turn" in 20th century philosophy.

---

[14] See Rudolf Carnap, *The Continuum of Inductive Methods* (Chicago: University of Chicago Press, 1952).

Carnap, then, accepts two distinct notions of probability. On the one hand, there is empirical or frequency probability based on actual counts of the frequencies of events or properties. On the other hand, there is logical or a priori probability which simply expresses symmetries of our descriptive language and the logical relations between evidence and hypotheses. The frequency assignments of probabilities are vulnerable to future experience and can be changed if experience requires it. The logical probability assignments are selected more or less conventionally and are invulnerable to anything which happens in nature. Insurance companies, who constantly revise their probability assignments on the basis of new experience, are using a frequency interpretation. Good dice players, who never fudge on the odds simply because of a long run of good or bad luck, are using a logical interpretation. Both, says Carnap, are equally valuable concepts of probability if we wish to understand how scientists operate.

A further function of the logical interpretation (to which we shall return later) is that of accounting for the use of the concept of probability in inductive reasoning. Neither Laplacian nor frequency theories of probability, it would appear, are capable of dealing with the notion. Thus, when we speak of a scientific hypothesis being made probable by certain evidence we mean to say neither (a) that its predictions have frequently come out right (and, by implication, sometimes wrong?) nor (b) that because of assumptions of equiprobability based on sufficient reason we are able to compute a high probability for the theory. Rather, we seem to mean something like "the logical relation between evidence e and hypothesis h lends high probability to the latter." In the sequel, this reading of the idea of confirmation will be seriously questioned.

PROBABILITY NOTIONS IN QUANTUM PHYSICS

But now let us ask: exactly how *do* scientists operate with probability concepts? How, for instance, do ideas of probability get involved in modern quantum physics? Basically, the answer is that the occurrence of statistical distributions and frequencies in experiments or empirical surveys forces scientists to fall back on probability. Conditions are produced which seem to be essentially constant (or variable only in a known way) and a variety of results are observed. All one can do is to count the number of times each kind of result crops up. Evidently the kind of probability involved here is relative frequency.

An example may help to make the point clearer.

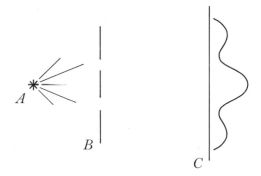

Suppose that an electron source (like the electron gun in a television tube) is placed in position *A* and a crystal— which we may suppose acts like a screen with two slits in it —is placed at *B*. A sensitized screen of some kind is placed at *C* to register the electrons as they strike. This is the

classic two-slit interference experiment.[15] When the apparatus is scaled suitably for sizes and distances the effect is a
rather striking one. Instead of electrons passing through
the slits and striking the screen in a fairly restricted region
(or pair of regions) an interference pattern is created. The
intensity of the scintillations—which indicates the number
of electrons striking at any one spot—is distributed as
shown on graph C. As physicists recognize, this pattern is
the familiar kind of pattern produced when a wave motion
interferes with itself. As a rough analogy, one can think of
water waves striking a breakwall B with passages in it. At
each entry into the calm harbor a disturbance will be set
up—a new wave—and any two such waves created will
run into or interfere with one another. Where their
troughs and crests cancel one another out little energy will
be transmitted to the beach C. Where crests coincide, a
large amount of water will run up the beach. The water-
marks on the sand might be expected to look something
like the graph of scintillation intensity on the figure.

From the present point of view, the important fact about
this experiment is that the results constitute a statistical
distribution. The graph of intensity can be thought of as
essentially similar to graphs used by social scientists to
show the distribution of characteristics like intelligence in
a sampling of people. The individual electrons, like individual persons, show up at different places on the chart.
Many of them turn up in the center and proportionately
fewer occur at the outer edges. Predictions about individuals must be given on a probabilistic basis here and we

---

[15] See, for instance, Niels Bohr, *Atomic Physics and Human
Knowledge* (New York: John Wiley and Sons, 1958), pp. 41–47.

can see from the diagram that an individual electron is more likely to turn up near the middle or one of the other modal points than in one of the trough areas. Here is where the probabilistic and statistical elements enter into quantum theory.

There is another matter of importance that should be brought out. If the emission of electrons at A is slowed down to the point where individual electrons leave at discrete intervals—an experimental feat of great delicacy —the remarkable result is that each transmits all of its energy to the scintillation screen at a single point (though not the same point in every case, of course).[16] Only as great numbers of them are sent from A to C does the interference pattern begin to show up. We can see, therefore, that the analogy of the water wave was an extremely gross one. An electron is not a wave. On the other hand, it is not an ordinary particle either; ordinary particles sent through the two slits would not form an interference pattern at all. This remarkable ambiguity in the nature of electrons and other fundamental "particles" is one of the most startling findings of modern science. It is usually referred to as the Complementarity or Wave-Particle Duality Principle and has become familiar to anyone with even a layman's interest in physics.[17] All of the fundamental entities making up the substratum of the observable world—including the pho-

---

[16] The experiment discussed here is described in R. M. Eisberg, *Fundamentals of Modern Physics* (New York: John Wiley and Sons, 1961), pp. 148–49.

[17] The idea of complementarity is due in the first instance to Niels Bohr. A relatively "un-metaphysical" treatment of it is given in Eugen Merzbacher's *Quantum Mechanics* (New York: John Wiley and Sons, 1961), pp. 4–9.

tons of light—share in this duality. To the extent that they do, statistical descriptions of their behavior are required.

In the elementary quantum theory, the state or condition of the electrons in our hypothetical experiment is denoted by a mathematical function ψ.[18] Using the laws of the quantum theory it is possible to deduce from ψ the probabilities for the various possible final positions electrons might go to. The mathematical details of the procedure are of less interest to us here than the following point: depending on the meaning or interpretation attached to the probabilities predicted by the theory a different meaning or interpretation will attach to the so-called state-function ψ. Its significance depends heavily on whether the probabilities we are dealing with are frequencies (as seems most plausible) or logical probabilities. I take pains to point this out because much of the discussion which has been carried on in the philosophy of physics concerning the meaning of the ψ-function in quantum theory has really been concerned with the question: what kind of probability does the quantum theory use? This has not always been recognized.

Let us try to elucidate the way in which probability interpretations affect the idea of the quantum state. It should become clearer as we go along that the frequency interpretation of quantum probabilities is not entirely without difficulties.

As we have already observed, one of the most distinctive differences between frequency and logical probability in-

---

[18] For a relatively non-technical exposition see Henry Margenau, *The Nature of Physical Reality* (New York: McGraw-Hill, 1950), pp. 337–38ff.

terpretations is that frequency generalizations seem to apply only to classes of events, not individual happenings. Logical probabilities, however, are capable of being used to speak about the probability of an individual event. When an insurance company determines on the basis of frequency data about deaths of 60 year old men that the frequency of death within the year between ages 60 and 61 is, say, .70, this number cannot legitimately be used in talking about the probabilities of the deaths of individuals. John Smith, age 60 and in excellent health, basking in the Florida sunshine, has far better prospects of making 61 than Sam Baker, a 60 year old cardiac case suffering in the smoggy atmosphere of Watts. It is ludicrous to say that the probability of each dying is .70.

Logical probabilities, however, can be applied to individual events. In Carnap's account of them, we simply need to formulate a hypothesis about an individual event—like getting heads on the toss of a coin—and then weigh the evidence. At the casino, one bets on the individual turn of the roulette wheel and it would be very strange indeed if one could not speak of the odds on hitting 13 on a particular spin for fear the real odds might change from one spin to the next.

Now if the probabilities of the quantum theory are truly frequency probabilities it seems to follow that they are probabilities describing the behavior of collections of electrons, not probabilities applying to the individuals. We cannot legitimately speak of "the probability that the next electron will strike the screen at position x" but only of "the frequency with which, in the long run, electrons will strike the screen at x." The implication for the interpretation of the state function ψ, in turn, is that ψ does *not* de-

scribe a property of individual electrons but only of collections of them. The individual electron does not literally
have a state ψ, it simply belongs to a collection whose characteristic or average behavior is nicely described by ψ.

We see here that what starts out as a nit-picking subtlety
about the meaning of the probability numbers suddenly
turns into an issue of deep ontological significance. If we
choose to interpret the probabilities predicted by the quantum theory as relative frequencies then the concept "state
of an individual electron" seems to have no real objective
content. To a physicist this might be tolerable. To philosophers of science—or, at least, to most of them—the idea of
a thing existing but not having a determinable state is a
metaphysical nightmare akin to Immanuel Kant's strange
concept of the *Ding an sich*. It is practically a tautology
that anything which exists must exist in some particular
state or describable condition. Exceptions are a bit difficult
to bring to mind.

The only plausible alternative to assuming that quantum probabilities are frequencies is to assume that they are
logical assignments of some kind. And yet this does not
appear to be entirely plausible either. For although logical
assignments do apply significantly to individual cases or
events (i.e., to individual electrons) they have the further
property of being independent of experience. A change in
our descriptive conventions or in our conventional choice
of an evidence-weighing rule can result in a totally different
set of probability assignments to the actual events dealt
with by the quantum theory. Surely this is an inappropriate
description of what actually goes on in quantum physics.
The grounds upon which probabilities are assigned to
electrons and the like are purely empirical. They consist
of data like the two-slit experiment, results of radioactive

decay processes and so forth. Methodologically, physicists find out what the probabilities are by making actual measurements and counts. They go through all of the same motions as insurance statisticians. How is this possible if the probabilities they deal with are really logical ones?

Subjective probability theorists may suggest at this point that the dilemma can be resolved by taking the probabilities of quantum theory as measures of the degree of belief or conviction physicists have about various predictions. This, however, must be rejected out of hand. The probabilities of the quantum theory, as we have seen, purport to deal with an objective randomness in the behavior of physical systems, not with the ignorance or knowledge level of the physicist himself. No, the probabilities of quantum theory—at least the basic ones—must be objective assignments. And that means they must be something like either frequency or logical probabilities.

THE NATURE OF STATISTICAL LAWS

What appears to be needed here is some clearer understanding of the nature of statistical laws and theories. This much is evident: general laws are the limiting case of statistical laws. They assert the existence of universally binding relations or connections between properties or things. Newton's law of gravitation is the philosopher's standard example: *all* pairs of mass points attract one another with a force inversely proportional to the square of their distances and directly proportional to the product of their masses. And by 'all' we mean '100 per cent of them.' If the percentage were something less than 100 we should call the law a statistical one, rather than a general one. That is the sense in which we say that general laws are the limiting case of statistical laws: they deal with regularities that occur not, as Aristotle used to say, "for the most part," but always.[19] The laws that tell us what happens "for the most part" (or even "for the least part") are statistical laws.

Because general laws are so closely linked, logically speaking, with statistical ones, anything ˙which holds for the former ought properly to hold also for the latter. Let us see where this leads us in the case of Arthur Pap's well-known thesis about the a priori character of some general laws and theories.[20]

Since the publication in 1946 of Pap's doctoral dissertation, *The A Priori in Physical Theory*, philosophers have been acutely aware of the fact that there is a certain invulnerability to disconfirmation or falsification exhibited by most of the general laws and general theories of the physical sciences. An excellent example of this is the relative invulnerability—Pap would call it the "functionally a priori" character—of the principle of conservation of energy. When in the early 1930's it was discovered that energy seemed to be lost from the nucleus of a radioactive atom in beta-decay, physicists refused to believe that the principle of conservation of energy had been disproved.[21] Its

---

[19] Aristotle, *Posterior Analytics*, II, 96 a 9–13: "Some events occur universally (for a given state or process may be true always and of every case), while others occur not always but usually; e.g., not every male human being grows hair on the chin, but it happens usually."

[20] Arthur Pap, *The A Priori in Physical Theory*, op. cit.

[21] In other contexts, of course, the Principle of Conservation of Energy has been seriously questioned by physicists. Bohr even favored rejection of the neutrino hypothesis for a time and was willing to alter the Principle! (See Wolfgang Pauli,

theoretical interconnections with other ideas in physics were so rich that it simply could not be abandoned lightly. Pauli, in a stroke of ingenious conservatism, proposed the hypothesis of the neutrino—a small, almost massless particle of neutral charge (and hence virtually undetectable) which carried off the lost energy by being emitted along with the beta particle.[22] More than 15 years were to pass before Pauli's conservatism was vindicated and the neutrino was detected experimentally. But few doubted during this time that something like the Pauli hypothesis had to be true: the principle of conservation of energy simply *could not* be found false.

Most of the basic principles and laws of classical physics have this character, as we saw in Chapter 3. Only in the face of the most massive and overwhelming evidence has any of them been overthrown or modified. Their logical interrelations imply that they cannot be abandoned one at a time, that a whole battery of principles must be altered at one time. You cannot, for instance, give up the principle of conservation of energy without calling into question Newton's third law of motion—the law that every action has an equal and opposite reaction. The Newtonian law of

---

Collected Scientific Papers, ed. by R. Kronig and V. Weisskopf, two volumes (New York: Interscience, 1964), Vol. I, pp. xvii–xviii, and Niels Bohr's letter to the editor in *Nature,* Vol. 138 (1936), pp. 25–26). It is obvious, therefore, that the Principle is not a synthetic a priori proposition in the tradition of Kantian idealist philosophy.

[22] J. F. Carlson and J. R. Oppenheimer report the proposal of the "neutron" (neutrino) in a letter to the editor, *Physical Review,* Ser. 2, Vol. 38, Part 2 (1931), pp. 1787–88.

gravitation cannot be given up without our reconsidering the concept of inertia. And so forth.

It needs to be emphasized that this logical feature of general laws in physics also holds for statistical theories and laws. There are some statistical laws which cannot be altered or given up without our getting involved in a drastic revision of basic physical concepts and theories. These statistical laws, I shall want to claim, involve something more than the ordinary frequency conception of probability and yet something less than the pure logical interpretation.

An illustration from the biological sciences will help us to understand how it is that theoretical support affects the vulnerability or invulnerability of a statistical law; how it is that some statistical generalizations become more than *mere* generalizations of data. Look for a moment at the elementary laws of heredity laid down by Mendel.[23] The simplest and most familiar is this: among the children of hybrid parents 1/4 will exhibit the pure dominant trait, 1/4 the pure recessive trait and 1/2 will themselves be hy-

---

[23] The fundamental laws of the Mendelian theory are as follows: "First, there are definite hereditary units . . . which are responsible for the transmission of characteristics. Second, there are two of each type of factor in the body cells of a mature organism. Third, when these two differ, one will be expressed (will be dominant) while the other will remain latent (will be recessive). Fourth, these factors segregate unchanged into the gametes so that each gamete carries only one factor of each kind. Fifth, there is a random union of gametes which results in a predictable ratio of characters in the offspring." A. M. Winchester, *Genetics* (Boston: Houghton-Mifflin, 1958), pp. 65–66.

brids. Because hybrids appear superficially to have the same characteristics as the dominants we say that $\frac{3}{4}$ will appear as dominants, $\frac{1}{4}$ as recessives.

In any actual sampling these figures are unlikely to be found. Only rarely does the law find exemplification and even then it is necessary to take very large samples. There is evidence that Mendel himself might have "dry-labbed" his observations in order to make them fit the statistical law.[24] Yet the law is regarded as almost beyond reproach. Geneticists would be astounded to find out that it is seriously defective.

The reason why this particular statistical generalization has such logical strength can be seen when one considers the elementary theoretical argument which supports it. Traits are passed on from parents to offspring by the random combination of gene pairs, one from each parent. If the parents are themselves hybrids their own gene pairs contain one dominant element and one recessive. The (equiprobable) combinations which can occur in the offspring are $D–D, D–R, R–D$ and $R–R$—two hybrids, a recessive and a dominant. *It is a simple point of logic that if Mendel's law is wrong some part of this theoretical argument must be wrong also.* Our whole conception of the process by which traits are passed on must be overhauled. That is why geneticists would be extremely surprised to find that Mendel's law had been disconfirmed.

Logically, the relation between theory and Mendel's law

---

[24] This controversial thesis is argued by L. C. Dunn in "Mendel, His Work and His Place in History," *Proceedings of the American Philosophical Society,* Vol. 109 (1965), pp. 189–198. See especially pp. 194–95.

is that of deductive implication: $\theta \to L$. Whenever a statistical law follows as a deductive consequence from a theory in this manner it is likely to take on a character of (relative) invulnerability. It becomes something more than the mere statistical generalizations which an insurance actuary employs in fixing insurance rates. The heart of the difference is that the actuary keeps on modifying his generalization as the evidence comes in. The scientist—like Mendel—puts his law into the archives. Not even a small adjustment of the probability numbers will be allowed without careful consideration of the implications for basic theory.

The relative invulnerability of certain theoretically supported statistical laws is closely linked to the question of how frequency generalizations can be applied to individual cases or small groups of cases. The relationship is something like this: the more invulnerable the generalization to changes and disconfirmation, the smaller the classes to which it can be meaningfully applied. *The more weight of evidence you need to alter the original generalization, the more applicable it will be to individuals.*

The point is nicely illustrated by dice games. Our expectations about the behavior of dice—like our expectations about mating of hybrids—involve a fundamental background of theory: in this case, theory about the equilibrium and symmetry characteristics of small solid cubes. (Let me emphasize parenthetically that it is a primitive physical theory we are talking about here and not some kind of vague principle of sufficient reason.) On the strength of this theory we assert that a well-made die, symmetrical and unweighted, will, when tossed in the usual random ways turn up an ace one-sixth of the time. It is almost (but not quite!) impossible to imagine this coming out wrong. According to Carnap's logical conception of probability it is logically impossible for it to come out wrong. The structure of the language we use to describe the dice tossing, together with our conventional assumptions about an inductive method specify that the probability must be 1/6. And yet if our theories about symmetry and equilibrium are incorrect it might very well be the case that 1/6 will not be the observed longrun frequency. This would be the outcome, for example, if as a matter of physical fact geometrically well-made cubes turned out always to have one side on which they could never lie in completely stable dynamical equilibrium. Because that theory is so well-established and reaches so far into the fabric of our other physical knowledge we do not expect this to happen, of course. *For that very reason the probabilistic laws of dice games can be applied to individual cases.*[25] If we were any less sure of the ideas of equilibrium and symmetry the possibility of individual tosses being "different" from one another would have to be considered and we should at least have to hesitate before saying that the probability of an ace on the next toss is 1/6.

Theoretical backing, then, distinguishes certain statistical generalizations as being functionally a priori—relatively invulnerable to disconfirmation. This, in turn, warrants applicability to prediction and explanation of individual cases. Such generalizations I shall refer to as 'semi-causal laws,' a label which seems to encapsulate the idea that they represent something more than mere summaries of data.

In terms of Carnap's two-probability notion what I am

---

[25] Other conditions must however be satisfied. See the discussion of reference class selection given below.

proposing comes down to this: the probabilities used in science are *all* basically frequency generalizations.[26] Some, however, have deductive support from more or less well-accepted physical, chemical or biological theories. These latter probabilities *seem* to be a priori assignments, but the appearance is largely deceptive. They are—to reiterate Pap's useful phrase—functionally a priori; which means that their invulnerability to empirical counterevidence is only skin deep. If the basic theory they are deduced from can be challenged then they will themselves be open to question.

Theoretical support, of course, involves more than just

---

[26] The same kind of point has been made by Professor Henry Margenau, *The Nature of Physical Reality, op. cit.,* p. 264: "The frequency theory supplies an epistemic definition of probability, the Laplacian formula a constitutive one." There is, in Professor Margenau's view, no incompatibility between these two definitions. It is as though one were to define 'length' in terms of certain measurement operations (epistemic definition) while also giving a mathematical definition of 'length' in terms of the abstract notion of spatial metric (constitutive definition). We have, then, one probability concept seen in two different aspects. Professor Margenau's formulation seems defective on the following grounds, however: it implies that every probability assignment can be construed as both a priori and statistical; but as we have pointed out there are clear-cut cases where the a priori element is not present (e.g., insurance statistics). It is also difficult to see how a constitutive definition and an epistemic one can both be used to arrive at probability numbers. A mathematical definition of 'length,' after all, does not tell us anything about the actual lengths of any particular physical objects. Clearly, Laplace's definition is as much an epistemic definition as the frequency definition.

deductive connection. A theory, as we have seen in Chapter 3, is a family of explanations, not simply an axiomatic schema. Deduction plays a role in its logical structure but other kinds of relationships are essential. *This is as much true of statistical theories as of non-statistical ones.* Statistical anomalies—distributions of events which contradict or appear to contradict those implied by our theories—shape the explanantia of statistical theories and give statistical law statements a large measure of their special semantical character. A semicausal (statistical) law is, like a general causal law, the product of its explanatory uses.

There is only one concept of probability used in science, then: relative frequency. But it occurs in two different ways. In contexts where theoretical background knowledge as well as bare data about frequencies is used to support a statistical law such a law will function as an a priori principle. It will be a semicausal law. But in theory-free contexts (as, for instance, in actuarial studies and some parts of the social and biological sciences) relative frequencies are constantly to be revised as new data about frequency come in.

In quantum physics statistical laws are of the type I have described as semicausal. The formalism of the state-function ψ serves as a basis for deducing statistical statements about distributions, probabilities and expectation values. All of these statements are interpretable as frequency generalizations. But they have something else as well: theoretical support. Because of that fact they are not immediately subject to alteration. If we find, for example, that experiment yields a different distribution of terminal electron positions than that depicted in the figure given earlier, we will not necessarily abandon our prediction. Instead we shall assume provisionally either that (i) we have not yet sent a large enough collection of electrons through the slits or (ii) that some significant factor has been overlooked in the construction of the experiment. If these provisional assumptions fail to "check out"—i.e., if a distribution pattern other than the expected one turns out to be replicable or repeatable and if none of the factors recognized in the quantum theory as being relevant to the explanation of the final positions turns out to be present—if all of this comes to pass, then we may be inclined to give up our original prediction as being false. But notice that this would require a serious job of surgery on the foundations of the quantum theory itself. Understandably, physicists do not take on such a chore without being absolutely sure that the surgery is necessary. After all, the history of physics is littered with examples of unneeded hypotheses proposed to explain away imagined difficulties.

STATISTICAL EXPLANATION AND COMPLETE EVIDENCE

The unified concept of scientific probability we have been discussing is essentially a relative frequency interpretation. Because of its distinction between theoretically supported and unsupported statistical claims, however, it might better be termed an Extended Frequency Theory or interpretation of probability. The main features of this view are the following: (a) probability assignments to a basic set of events can be made in two ways, both ultimately based on observed relative frequencies but one only indirectly so. It is a truism to say that both types of assignment are used in science and that they are used interchangeably. But (b) the two forms of assignment are distinguishable and distinct. Assignments made on the basis of accepted physical, biological or other theories can and do disagree with observed frequencies. There is no logical necessity of their agreeing or even converging in the limit to a common value. (c) Probabilities assigned by direct observation of frequencies are to be revised at every new trial; probabilities assigned by derivation from fundamental theory are revised only if the new data suffices independently to falsify the fundamental theory itself.

We are not, of course, contradicting the usual position of the frequentist but, rather, offering an elaboration of it aimed at showing that the work of logical probabilities can be done by the frequency theory.

Like any frequency interpretation, an Extended Frequency Interpretation is open to criticism and challenge. In this section and the next, therefore, I should like to consider two of the most important of these objections. The first charges that the frequency theory of probability is inherently inadequate to deal with the job of analyzing statistical explanation in the natural sciences. The second objection charges that even if statistical explanation were analyzable in terms of frequency concepts one would still require some species of logical probability to deal with the ideas of induction and confirmation in science. Let us consider the matter of statistical explanation first.

Statistical explanations come in two shapes. Hempel calls them "Inductive-Statistical" and "Deductive-Statistical." [27] As an illustration of each, recall the two-slit experiment with electrons described earlier. Suppose our expectation before observing the outcome of the experiment is that the

---

[27] C. G. Hempel, *Aspects of Scientific Explanation, op. cit.,* pp. 380–81.

electrons ought to cluster about two points on the scintillation screen. We believe, for example, that electrons are really particulate in nature and hence are convinced that the electron beams ought to behave like trains of little billiard balls. The observed distribution, then, is anomalous with respect to our antecedent beliefs.[28] We require an explanation.

A quantum theoretician explains the distribution to us by showing us that in the context of the elementary quantum theory—wherein wave-particle duality is assumed— the resultant spread of positions is a deducible consequence. The statistical (and nonstatistical) laws of the theory are employed to deduce a (statistical) explanandum statement describing a "fair sample" of electron positions. This is the kind of explanation Hempel terms Deductive-Statistical.

Suppose, on the other hand, that we had originally expected some sort of distribution of the kind observed but were convinced (for whatever reasons) that a particular electron—call it Alpha—ought to have shown up in a position different from its actual landing place. In this instance our expectations concerned the position of a single electron, not the entire collection of them. The anomaly is singular rather than plural.

The quantum theoretician can still explain the result but he must now do something a little different. He must infer from his theory some non-zero probability for the event of Alpha landing at the position actually observed. He needn't show us that the distribution we have observed

---

[28] The statistical character of this and similar anomalies is discussed in the final section of the present chapter.

is understandable. He only needs to illustrate how Alpha's behavior is compatible with a proper understanding of what electrons are and what they do. (Some philosophers might add that Alpha's appearing at the specified position must be shown to be "more probable than not." This seems an unwarranted restriction on statistical explanation of individual events and will not be included here. It is difficult to see why the requirement is even thought to be reasonable; statistical theories must surely be capable of explaining improbable events.)

In this second form of explanation—explanation of the Inductive-Statistical form, Hempel calls it—individual events are treated statistically. It is at this point that the alleged impossibility of carrying through a consistent frequency interpretation of probability arises. In order to see why that may be so we need to look more closely at the logical form of the two sample explanations.

The logical character of Inductive- and Deductive-Statistical explanations depends basically on the nature of the explanandum statement involved. If the explanandum is itself a statement about frequencies—a percentage or distribution statement, say—then the pattern of explanation is no different, in principle, from that of the deductive-nomological explanation. Statistical laws will be adduced in the explanans but since the explanandum is itself statistical the logical relation between the two is simply the ordinary relation of deductive connection. The quantum mechanical explanation of the total distribution of electrons fired in the two-slit experiment was our example of this form of explanation. In Chapter 6 a number of further concrete instances will arise when we discuss the history of the quantum theory of mesons.

The format for this first type of statistical explanation

(Deductive-Statistical) is essentially the same as the pattern discussed in Chapter 2 (Deductive-Nomological). Statistical laws $S_1, S_2, \ldots, S_n$ appear in the explanans (with perhaps a few nonstatistical laws included). In most cases descriptions of relevant empirical conditions $C_1, C_2, \ldots, C_m$ are added. The explanandum statement E—describing an observed frequency, distribution, average or the like follows deductively from these.

$$S_1, \; S_2, \ldots, \; S_n$$
$$C_1, \; C_2, \ldots, \; C_m$$
$$\overline{}$$
$$E.$$

An anomaly context, of course, is implicitly involved. It may be either statistical in character or not.

Most of the statistical explanations offered by modern physics are of the statistical-deductive type. Scientists, generally speaking, are interested in explaining ensembles of events, not individual happenings. Yet there is a sense in which statistical laws do apply to individual facts and events. Quantum mechanics says *something* illuminating about the individual electrons in the two-slit experiment, albeit something difficult to spell out in logical terms. In the previous section we sought to show that the capacity of a statistical law to apply to individual events was a function of the theoretical backing enjoyed by the law. Now we must say something more.

The problem about Inductive-Statistical explanation of individual events according to some philosophers (notably Hempel) is that the relation between explanans and explanandum is not a deductive one. According to this view, it is impossible in principle literally to deduce anything from a statistical law concerning the probability of occurrence of single events. *Hence the inference must be considered an inductive one.* The upshot is that one must employ inductive (or a priori) probability in order to characterize the relation between explanans and explanandum. The relative frequency notion of probability must be augmented by the a priori (inductive) concept.

Hempel's description of the pattern of Inductive-Statistical explanation is instructive in this regard. According to Hempel, we infer only an inductive probability statement from explanatory premises of the form '$f(A, B) = f_1$' (read: the frequency with which Bs are also As is f_1—a number between 0 and 1) and 'the individual Alpha belongs to the class B.' The statement that we can infer asserts that the degree of confirmation (Carnap's $C(h, e)$ discussed earlier) for the hypothesis 'Alpha is also in A' is exactly f_1. Schematically,

$$f(A, B) = f_1$$
Alpha is a B
––––––––––––––––– [Probability: f_1]
Therefore, Alpha is an A.

In part, the presumed need for inductive probability in this schema rests on a misconception about the difference between genuine statistical laws (semicausal laws) and mere statistical generalizations. Unless '$f(A, B) = f_1$' is a statistical law with theoretical backing the inference set out above will not generally be a valid one. Introducing inductive probabilities does nothing to mitigate the evil. Consider, for instance, the example suggested earlier. If '$f(A, B) = f_1$' is a statement about the frequency of death of 60 year old males and if 'Alpha' stands ambiguously for John Smith, a retired Floridian in excellent health, or Sam Baker, a 60 year old cardiac case in Watts, the inference outlined above is obviously incorrect. We cannot ascribe the same probability of death to both men. Calling the probability 'inductive' or 'logical' or 'a priori' does nothing to lift this restriction. *Unless the frequency statement appearing in the premises has a background of theoretical support nothing can be inferred about individual cases.*

If we introduce the Extended Frequency interpretation at this point, therefore, a large measure of the difficulty disappears. The inference from explanans to explanandum *can* be construed as a rigorous deductive one simply by introducing the inference rule that the probability of an individual event being f_1 follows from a statement about frequency if and only if the frequency statement is a semicausal law (i.e., has theoretical support within some well-established body of knowledge).

The demand for an inductive account of statistical explanation, however, does not rest merely on a misconception about the nature of statistical laws. Even if semicausal laws are distinguished from mere generalizations and even if it is agreed that the explanans of a statistical explanation must include such laws—even if all this is granted, a serious conceptual problem remains. Mere subsumption under a statistical law does not necessarily count as explanation (any more than mere deduction from general laws does!). We must find the *right* statistical law to apply to the individual case. We must, in other words, put the individual case in the right reference class in order to explain it.

The problem of the reference class can be illustrated nicely by means of the so-called "paradox (or ambiguity) of statistical explanation." [29] The argument runs as follows:

---

[29] C. G. Hempel, "Deductive-Nomological vs. Statistical Explanation" in *Minnesota Studies in the Philosophy of Science,* Vol. 3, *op. cit.,* pp. 125–28ff.

Suppose that '$f(A, B) = f_1$' is a statistical law and that Alpha is a B. We infer that the probability of Alpha being in A is likewise f_1. Let C now be a class formed from the Bs and including exactly 10 elements: Alpha and 9 of the Bs which are known not to be in A. Since we are trying to explain statistically why Alpha is an A, it is apparent that we know Alpha *is* an A. Hence, $f(A, C) = .10$. We assume for the sake of argument that this is a statistical law. But then from '$f(A, C) = .10$' and 'Alpha is in C' it can be inferred immediately that the probability of Alpha being in A is .10. Since f_1 was arbitrary to begin with, we have obviously arrived at a contradiction. By suitable construction of the artificial reference class C, *any* probability can be given to the statement 'Alpha is a member of A.' Clearly, some restrictions on the admissible classes C are needed.

The problem can be stated more generally and more simply. Suppose we are trying to explain a particular fact or occurrence. We know that an existing theory supplies us with several statistical laws relating to the occurrence. Depending upon how we classify the fact or event we shall have to use a different set of statistical laws (and hence arrive at different estimates of the likelihood that the event should have occurred). How do we choose the proper reference class?

A single electron strikes position p on the scintillation screen in the double-slit experiment. Suppose we have two sets of information about the electron. The first body of information states that this electron belongs to an ensemble of electrons in state ψ, all of which hit the screen today. The second body of information tells us, say, that this electron belongs to a combined collection consisting of electrons fired today *plus* all electrons fired in a single-slit experiment yesterday. The frequency with which electrons hit p

in the wider class is manifestly different from the observed frequency as measured today only. Both probabilities are given by statistical laws of the quantum theory. We *know* that the probability based on classifying the event as one of today's outcomes is the right one. We *know* that it is a conceptual error to include yesterday's results in our discussion of this particular case. But *how* do we know this? What foundation does our assumption have?

The seriousness of the problem of the reference class now becomes more evident. Even on the assumption of the Extended Frequency Theory of probability we find that contradictory probability assignments follow from acceptable semicausal laws. *Inference from a semicausal law to the particular case cannot therefore be construed as valid deduction unless some further restriction is introduced.*

The classical solution to this puzzle among frequentists is the so-called Rule of Complete Evidence. As formulated by Hans Reichenbach, this rule asserts that when two classes *B* and *C* are both known to contain the individual *a* then the appropriate class for explaining or predicting things about *a* is the intersection of *B* and *C* (i.e., the set containing events common to both *B* and *C*).[30] The inter-

---

[30] Hans Reichenbach, *The Theory of Probability* (Berkeley: University of California Press, 1949), p. 374: "We . . . proceed by considering *the narrowest class for which reliable statistics can be compiled.* If we are confronted by two overlapping classes, we shall choose their common class. Thus, if a man is 21 years old and has tuberculosis, we shall regard the class of persons of 21 who have tuberculosis. Classes that are known to be irrelevant for the statistical result may be disregarded. A class *C* is irrelevant with respect to the reference class *A* and

section of the sets of electron hits yesterday and today with the set of results from today only is just the set of today's results. Hence that is the proper reference class. Any information which delimits the class of objects to which a is referred must be included.

As customarily stated the Rule of Complete Evidence suffers from the defect that irrelevant as well as relevant factors must be used in selecting the reference class. Thus, if it is known that Sam Baker—our 60 year old cardiac case in Watts—has four toes on his left foot, and if the frequency with which Watts' four-toed 60 year old cardiac cases die is different from the frequency of death of five-toed folk in the same condition, we must utilize Sam's four-toedness in explaining his death. Causally speaking, one knows that the missing toe had nothing to do with it. But the Rule of Complete Evidence does not appear to permit us to recognize that fact.

Struck by the seeming inadequacy of the Rule of Complete Evidence, some logicians have sought to formulate new kinds of criteria for delimiting the possible reference classes which may be used in statistical explanation. The idea is to develop some means of excluding classes which are too broad while at the same time not going to the oppo-

---

the attribute class B if the transition to the common class A. C does not change the probability . . . For instance, the class of persons having the same initials is irrelevant for the life expectation of a person." Reichenbach concedes (p. 375) that this method is not "perfectly unambiguous. Sometimes it may be questioned whether a transition to a narrower class is advisable." This is dismissed as a pragmatic point. It is also noted that if the frequency of property B in the common class A. C is unknown no inference can be drawn.

site extreme of choosing overly narrow classes. The rules proposed are generally rather complicated. One of the most recent—that of C. G. Hempel—even turns out on close analysis to rule out *all* classes whatever as reference classes! [31]

---

[31] Hempel, *Aspects of Scientific Explanation, op. cit.,* pp. 399–400: "[The Requirement of Maximal Specificity for Inductive-Statistical Explanations.] Consider a proposed explanation of the basic statistical form

$$f(G, F) = r$$
$$Fb$$
$$\overline{} \quad [\text{Probability: } r]$$
$$Gb.$$

Let s be the conjunction of the premises, and, if K is the set of statements accepted at the given time, let k be a sentence that is logically equivalent to K (in the sense that k is implied by K and in turn implies every sentence in K). Then to be rationally acceptable in the knowledge situation represented by K, the proposed explanation must meet the following condition (the requirement of maximal specificity): If $s \cdot k$ implies that b belongs to a class F_1, and that F_1 is a subclass of F, then $s \cdot k$ must also imply a statement specifying the statistical probability of G in F_1, say

$$f(G, F_1) = r_1.$$

Here, r_1 must equal r unless the probability statement just cited is simply a theorem of mathematical probability theory." Hempel's requirement, as we said, rules out any class whatever. This can be shown as follows: If r is not a rational number then the frequency of Gs in any *finite* class F_1 must be different from r. Hence if background knowledge includes information to the effect that b and at least one other object

The truth is that no new rule is needed at all. The Rule of Complete Evidence is entirely adequate—provided that we recognize a single limitation. It is the following: *A new piece of evidence will be relevant to our predictions and explanations of Alpha's status only if it is connected to Alpha's status by a genuine statistical law.* We must have at our disposal some reasonably well-established theory about connections between the factor and the status of Alpha before the factor can be counted. If an individual having the dominant genetic trait D is born of hybrid parents and if he happens to be the second offspring in the family we shall count the latter fact as relevant only if genetic theory tells us something definite about second offspring and dominant traits. In point of fact, genetic theory tells us no such thing. It implies no unique difference in characteristics for second born. Even if it is found empirically that second born children tend to inherit the dominant trait more frequently than others we would not be obliged to include mention of the characteristic. Only direct information about the nature of the genes and chromosomes themselves could make a difference here. We should have to learn that there is some alteration in genes, chromosomes

---

are members of F we form the finite subclass of F including those members and call it F_1. If, on the other hand, r is a rational number—call it p/q—we simply select the subclass F_1 by taking b along with any other n elements of F, $(n < q + 1)$. The frequency of Gs in F_1 then cannot equal r. Since background knowledge in any actual case always implies something about the membership of the given class F there will never be any problem about constructing the subclass F_1 according to these rules and Hempel's requirement can never be satisfied.

or reproductive process between first and second offspring before we should be willing to allow that the fact of being second born is relevant to dominance.

A similar example can be imaged in quantum theory. We might find, for instance, that electrons seem to pile up at p more rapidly during the later stages of our experiments than during the early stages. In this case, it could be asked whether the fact that p was a "last half" or "first half" electron should be included in the selection of a reference class. Only if some aspect of the state of the electron could be related to the "first half-last half" distinction, though, would this fact become relevant. And that, incidentally, would require a complete overhaul of the quantum theory as now understood. (See the discussion of von Neumann's theorem in Chapter 5.)

In so-called Inductive-Statistical explanation, then, the selection of a reference class follows the rule that all evidence concerning the law-like statistical relationships of the individual to other members of the class must be taken into account. The reference class should be the "smallest" set to which the individual is known to belong, save that its "accidental" memberships in subclasses can be overlooked. When the "smallest" such set is employed, inference from the statistical generalization to the particular case is a matter of course. It is deductive, and, more important, it is explanatorily relevant.

THE IDEA OF INDUCTIVE LOGIC

There remains a second important objection which might be lodged against an extended frequency interpretation by defenders of the two-probability doctrine. It runs as follows:

You claim that physical or biological theories can support statistical laws and render them functionally a priori. But those theories, on your own admission, must be 'well-established' or 'well-confirmed.' In plainer language, the theories must have some degree of probability on the basis of empirical evidence. *This is nothing other than the idea of degree of confirmation or logical probability.* Even if you do eliminate the use of logical probability from scientists' explanations and predictions about phenomena, therefore, you do so only by assuming that the theories they use can be described as probable in the sense of being 'well-confirmed.' You do not, then, succeed in eliminating logical probabilities from science, you only restrict their application to estimates of the probability of whole theories.

The rejoinder to this argument is that it is unnecessary to use the idea of logical probability in order to analyze the idea of "well-established theory." Indeed, any attempt to attach probability numbers to theories like evolution, special relativity, or elementary quantum mechanics seems ill-conceived right from the start. To put it bluntly: there is no such thing as inductive logic (in the sense of a probability calculus for evaluating the probability of hypotheses and theories).

This thesis is not novel. Karl Popper and other philosophers of science have argued for many years that belief in an inductive logic is an illusion.[32] According to Popper,

---

[32] Karl Popper, *The Logic of Scientific Discovery* (New York: Basic Books, 1949), p. 29: "My own view is that the various difficulties of inductive logic . . . are insurmountable. So also,

theories are corroborated by being put through tests; the severity of the tests passed determines whether the theory becomes "well-established" or not. There is no inductive inference from evidence to hypotheses; rather, there are series of deductive inferences from hypothesis to predictions and the success of these predictions—particularly the ones relating to crucial tests—increases our confidence in the hypothesis.

The argument I wish to offer goes beyond Popper's proposal in two ways. (1) I wish to argue that the logical relations between data statements and general theories actually relevant to the notion of "well-confirmed theory" are deductive, rather than inductive. (2) I want to maintain that 'confirmation' is a systematically ambiguous term covering very different relationships and hence one not amenable to the unified treatment envisioned in the logical theory of probability. If these theses are accepted it follows that it is neither necessary nor even possible to analyze confirmation in terms of the probability calculus. This applies not only to the Carnapian concept of degree of confirmation but to other forms of inductive logic as well.

Most logicians wince at assertions like (1). Everyone knows, after all, that you can't literally deduce a general law or set of laws from experimental or observational data.[33] Deduction never goes from the particular to the

---

I fear are those inherent in the doctrine, so widely current today, that inductive inference, although not 'strictly valid,' *can attain some degree of 'reliability' or of 'probability.'* "

[33] Stephen Toulmin, *The Philosophy of Science, op. cit.,* p. 41: ". . . it is essential to see at the outset that there can be no question of observation-reports and theoretical doctrines being connected [deductively]. . . . This comes out clearly from our

general, it is sometimes said. Observation never logically determines which theory we must accept (and, in general, this is true). These claims are only really relevant, however, if we presuppose that no background of theory is involved when a scientist generalizes his laboratory results. On the contrary: some background knowledge of a general character is *always* assumed in the laboratory. And against a background of general assumptions it *is* possible literally to deduce laws and theories "from the phenomena" (as Newton used to say).[34]

For instance: electrons are all assumed by the quantum theory to be basically identical. They differ only in regard to a few key properties—position, velocity, and so forth. An experiment with a single electron can, therefore, be sufficient for deducing general conclusions about all elec-

---

example: however many statements you collect of the form, 'When the sun was at 30° and the wall six feet high, the shadow was ten feet six inches deep,' you will not be able to demonstrate from them in a deductive manner the necessity of the conclusion, 'Ergo, light travels in straight lines.' " Toulmin, of course, is looking at the wrong kind of experimental evidence. The point is, though, that he does believe one cannot literally deduce anything from experimental data. An effective and provocative rebuttal of that position is given in Victor Kraft's essay "The Problem of Induction" in *Mind, Matter and Method: Essays in Philosophy and Science in Honor of Herbert Feigl*, ed. by P. K. Feyerabend and Grover Maxwell (Minneapolis: University of Minnesota Press, 1966), pp. 306–318.

[34] Newton, *Principia, op. cit.*, p. 547: ". . . I frame no hypotheses; for whatever is not deduced from the phenomena is to be called an hypothesis . . ."

trons. The conclusions, of course, must not include or presuppose information about the properties in which electrons differ from one another. But to the extent that we
know how similar one electron is to another we are warranted in drawing general conclusions from particular
experiments. What looks like inductive reasoning here
is really analyzable as deductive. *Where we know of de
facto regularities in nature, where we have analogies and
models to fall back on, we can infer from the particular
to the general with confidence.*

Nor is the type of generalization we can make restricted
to universal generalization. Statistical results also follow
from the essential similarity of any large or random sampling of electrons to any other such sampling. Electron
diffraction observed in one experiment can be freely and
deductively generalized to any similar experiment because
of this fact. John Stuart Mill made that point over a century ago.[35] He spoiled it, however, by going on to assert

---

[35] John Stuart Mill, *A System of Logic: Ratiocinative and
Inductive* (New York: Harper and Brothers, 1860), p. 185:
". . . every induction may be thrown into the form of a
syllogism, by supplying a major premiss. If this be actually
done, the principle . . . of the uniformity of the course of
nature, will appear as the ultimate major premiss of all in-
ductions . . ." This is explained further in a footnote where
Mill argues against Archbishop Whately that it will not do
to assume only a *specific* generalization in support of an inductive inference since that generalization would require to be
proved inductively. "It hence appears, that if we throw the
whole course of any inductive argument into a series of syl-
logisms, we shall arrive by more or fewer steps at an ultimate

that there is some overarching general principle of the Uniformity of Nature sufficient to render any generalizing argument whatever deductively valid. No such claim will be made here. There is no all-encompassing Uniformity of Nature, there are only uniformities. And there *are* such things as bad generalizing arguments.

The kind of generalizing inference we are discussing can best be described as *generalization by second-order laws or hypotheses*. The phrase 'second-order' refers to the fact that such laws and hypotheses involve quantification over predicates and relations. Symbolically, they have the form

(A) $(x) [Fx \supset (G) (Gx \,\&\, \phi G \supset (y) (Fy \supset Gy))]$

where ϕ restricts the type of property or relation G must be in order to be generalizable. Such a proposition asserts that any property G meeting the general condition ϕ can be ascribed to all Fs if it can be ascribed to any individual F. Clearly, it is possible to infer deductively the general law '$(y) (Fy \supset Gy)$' for particular G, from (A) and premises of the type

---

syllogism, which will have for its major premiss the principle, or axiom, of the uniformity of the course of nature." This, of course, is a non sequitur even for elementary syllogistic logic. Remarkably, Mill goes on to say (p. 189) "In the contemplation of that uniformity in the course of nature which is assumed in every inference from experience, one of the first observations that present themselves is, that the uniformity in question is not properly uniformity, but uniformities. The general regularity results from the coexistence of partial regularities."

$$\begin{cases} \phi G \\ Fa \ \& \ Ga \end{cases}$$

where a is some individual case.

The reason inductive logicians reject such inferences as being significant for the theory of confirmation is that they assume a circularity is involved in allowing (A) to constitute part of the evidence for the conclusion '(y) $(Fy \supset Gy)$'. According to this line of argument, (A) can be confirmed only by looking at instances of F and G, and *a fortiori*, by seeking empirical confirmation for '(y) $(Fy \supset Gy)$'. Hence '(y) $(Fy \supset Gy)$' counts as evidence for (A), not the reverse.

This argument hinges, however, on an ambiguity in the term 'confirmation.' In one sense—the crucial sense for purposes of inductive logic—to confirm is to provide empirical proof, a kind of proof analogous to, but in some way looser than, mathematical proof. In an entirely different sense, 'to confirm' is merely to verify a proposition as holding true over some specified range of cases. Propositions like (A) can often be confirmed in this latter sense. We can, for example, verify that various properties of electrons, and electrons themselves have always behaved in accordance with some principle like (A). *This in no way says, however, that (A) can be or has been confirmed in the more fundamental sense.* Indeed (A) is the kind of proposition which generally speaking *cannot* be confirmed in the strong sense. It remains always a working *hypothesis* subject to disproof by single counterinstance. Beyond saying that it has not so far been violated by experience there is nothing positive we can say in its defense.

The second type of confirmation mentioned above I shall refer to as "Instance Confirmation." The first and more fundamental type may be termed "Second-order Confirma-

tion" or "Confirmation by Second-order Generalization."
The distinction between the two is essential to our critique
of the inductive logic (or logical probability) program. For
if it is granted that generalizations can be supported in
these two distinct ways it follows that the attempt to define
a unitary concept of confirmation (such as C (h, e) represents) is doomed to failure. Exactly the same body of *empirical* data statements (e.g., '*Fa*' and '*Ga*') will support a
general claim differently depending on whether we are
speaking about instance confirmation or second-order confirmation.

To make this manifest look at any body of empirical
data relevant to either

(A) $(x) [Fx \supset (G) (Gx \,\&\, \phi G \supset (y) (Fy \supset Gy))]$

or

(B) $(y) (Fy \supset Gy)$, for particular G.

Then ask to what extent the data supports (B) on the assumption of (A) and to what extent the same data supports
(B) if (A) is known to be false. Information about the
ϕ-hood of G or about other properties G' of the Fs will
very likely be irrelevant if (A) is false. The "degree of confirmation" of (B) by given evidence e, therefore depends
intrinsically on the truth or falsity of certain background
assumptions and cannot be construed as resting solely on
logical relations between e and the generalization alone.

Broadly speaking, the general statements of empirical
science divide into two categories: hypotheses or conjectures (as Popper would call them) for which conclusive
empirical evidence is unavailable, and generalizations

whose truth is affirmed by empirical evidence. The hypotheses—e.g., that two electrons are indistinguishable except for position, velocity, etc.—form the background of knowledge against which generalizations are drawn. We *deduce* —literally—generalizations using these very general hypotheses. In some cases, the hypotheses themselves come to have empirical support. As this happens, the body of theory into which they are embroidered becomes stronger and more cohesive. The basic pattern of confirmation is the deductive pattern of second-order generalization. Instance confirmation plays a secondary and subordinate role.

To say that a theory is "well-confirmed" or "well-established" is not to say that it is proved. Any theory in science involves hypotheses in some capacity or other and therefore must be thought of as open-ended. (An empirical theory which can literally be said to be experimentally proved is one whose generalizations are so interrelated that each can be deduced from empirical data statements using the others as background assumptions. It is obvious—or should be—that no such theories exist in the natural sciences.) If, however, there are no known anomalies with respect to the hypotheses of the theory and if at least some of the general claims of the theory are deductively supported by empirical data statements then we can speak of it as "confirmed" or "well-evidenced" on the whole. If more of the theory's generalizations come to have empirical support one can say that the "weight of evidence" for the theory has increased. (It might even be possible in some cases— where, for instance, the number and kinds of hypotheses involved are directly comparable—for one to say that theory θ_1 is better supported by evidence than θ_2.) But the main thing is that formal notions of probability are just not involved.

Generalization, of course, is not the only form of inductive or pseudo-inductive inference and we close this section with brief mention of another kind of inference sometimes thought to require an inductive logic for its evaluation. This kind of inference is usually termed "eliminative induction" and concerns a process by which hypotheses may be defended as well-founded.

Eliminative induction, as the title suggests, proceeds by the elimination of alternative hypotheses until only one compatible with the evidence remains. The implicit assumption, of course, is that all *possible* hypotheses have been considered. And when that assumption is made the eliminative inference simply turns out to be a deductive one. Beginning with the disjunction of the various hypotheses we deny one after another until a single one is inferred.

The trouble with this type of inference, both practically and logically, is in knowing that the disjunction is complete—that *all* hypotheses are included. Because there are, in principle, an infinity of hypotheses which might be considered logically possible and relevant it would appear that the needed disjunction can never be asserted. Some proponents of logical probability regard this as a justification for invoking their particular theory of confirmation to explicate the whole business of eliminative induction.

What must be borne in mind, however, is that elimination normally is employed in the context of some more or less well-developed theory and against a background of simplifying hypotheses (e.g., "No force in nature requires differential equations of order greater than four for its representation"). As we saw in the work of Leverrier and Newton in Chapter 1 it sometimes is possible to restrict alternative explanations to a finite number *within* a given theory. The procedure is that of defining an anomaly by

showing the logically possible kinds of hypotheses allowed
by the theory at hand.

Eliminative induction, then, can be considered as deductively structured when theoretical background is assumed.
When theoretical background is unavailable the technique
seems not to be usable at all. (In any case, no examples
of valid eliminative induction exist which are theory-free.)
Eliminative induction, therefore, does not require a logical
theory of probability. In fact, as we have seen, nothing in
the logic of confirmation does. Confirmation occurs basically through deductive logical connections between data
and generalizations. Instance confirmation, where it occurs,
plays a secondary role. The attempt to analyze confirmation
in terms of logical probability, therefore, seems neither
necessary nor possible.

STATISTICAL ANOMALIES

Statistical explanation, like causal explanation, involves
implicit appeal to an anomaly context. Sometimes, though
not always, the anomaly context will itself be statistical
in character. One or more of its statements may be statistical laws or generalizations. It is essential to our understanding of explanation in quantum theory, genetics, and
similar theories to see how anomaly contexts of a statistical
type are structured.

In order to round out our discussion of statistical theories, therefore, we shall seek in this section to analyze the
concept of "statistical anomaly."

Basically, there are two types of statistical explanations
—explanations of distributions, frequencies or averages,
and explanations of individual happenings. *But there is
only one kind of statistical anomaly. Only distributions,*

frequencies and averages can be anomalous with respect to statistical laws. Individual events cannot. The description of an individual event simply cannot contradict a statistical generalization.[36] It can, of course, contradict the non-statistical portion of a theory underlying a statistical generalization. This is the case if, for instance, a particle should land completely outside the range of predicted positions laid down by the quantum theory. A frequency statement or a statement of average value, however, cannot be contradicted in this fashion.

An individual event can, of course, be improbable (in the sense of having a low predicted probability from known statistical laws). But improbability is not the same thing as anomalousness. Getting 10 straight deuces at dice is highly improbable. But it is not anomalous—unless one holds the (mistaken) belief that highly improbable events never occur. No special explanation of the 10 deuces is required so long as we understand that such a run is not ruled out by the probabilistic theory of dice games.[37]

---

[36] We exclude the cases of frequency generalizations in which the frequency is either 0 or 1.

[37] This point is one which needs to be reiterated in view of the specious arguments often heard concerning so-called extrasensory perception. That a person predicts the turn of cards more frequently than the "laws of chance" provide does *not* prove that something is unusual about the person. His behavior is entirely compatible with the usual assignments of probabilities to turns of the cards and, what is more important, it does not contradict any psychological theory about the probability of a person's predicting correctly. The only thing it does contradict is the naive and somewhat foolish belief that highly improbable events never occur.

The confusion of anomalousness with improbability is not easy to avoid. To say that an event is anomalous is, in ordinary English, to say that the event is unusual or surprising. And that has much the same significance in common speech as saying that the event is one we regard as improbable. The use of the word 'probable' or 'improbable' in this context, though, is different from its use in the mathematical theory of probability. The probability of getting a deuce on a pair of dice is 1/36, a rather small number compared, say, to ½. Getting a deuce is therefore relatively improbable. But it is neither unusual, surprising, nor anomalous for deuce to come up in dice games. We expect it as a matter of course. A person who demanded an explanation why deuce came up, saying "The probability of that event is extremely low," would obviously be confused in his reasoning. Improbability as such does not give us grounds for a reasonable request for an explanation.

Statistical theories in science normally seek to explain distributions, not individual events. The quantum theory of mesons, as we shall see, is occupied with the task of rationalizing collections of data about the average values and distributions of values of cosmic ray energies and the like. Rarely does the problem arise of explaining an individual event via a statistical theory. When that problem does come up it comes up because the individual event seems to contradict non-statistical expectations. Mere improbability of individual events is not deemed sufficient reason for seeking an explanation. Thus, a single heavy track in a cloud chamber may be construed by quantum theorists as anomalous with respect to their theory. But it is only seen as such because it appears to violate non-statistical assumptions about the energy that a cosmic ray may have. No statistical law of the quantum theory is challenged by such an event, only the non-statistical foundations underlying all of the statistical laws.

A frequency or distribution can be anomalous with respect to a statistical law if it contradicts that law. Establishing that such a contradiction exists is often difficult and laborious. One must, to begin with, look at a fair sample or large random selection of cases in order to formulate the frequency or distribution statement. By "fair sample" we mean "a set of cases sufficient for purposes of generalization." In other words, a fair sample is sufficient for drawing a (deductive!) inference to some general conclusion. It is obvious that the nature of a fair sample depends upon the subject matter within which one is working. A fair sample of fruit flies is numerically (and generically) different from a fair sample of pendulums. The regularities we have already established or hypothesized about the subject matter form the logical basis for our inferences to general conclusions. Statistical anomalies, then, are established by (a) examining a fair sample of data, (b) generalizing the sample data by means of deductive reasoning—a process which requires general background knowledge, and (c) showing that the generalizations —statements of frequency or distribution—are incompatible with those generable from the known statistical law.

5

Applicability, Scope, and Extension

As we saw in Chapter 1, Einstein's explanation of the photoelectric effect raised serious doubts about the traditional picture of light as a train of electromagnetic waves. The transmission of energy to the photoelectron does not proceed continuously—as the wave picture dictates—but involves discrete pulses more characteristic of particle impacts. This fact, together with results obtained by later workers showing that so-called elementary particles (e.g., electrons) exhibit periodicities characteristic of wave motions, forms the basis for the present day quantum theory. Extraordinary results like the two-slit experiment discussed in Chapter 4 all but clinch the case.

The introduction of wave-particle duality for both light and "elementary particles" has not gone unresisted. It has provoked much (if not most) of the controversy and perplexity experienced by philosophers of science in the 20th century. Einstein himself objected strenuously and sought other ways of interpreting the experimental data. More recently, Professor David Bohm has taken up the cudgels in an attempt to restore a clear-cut distinction between wave and particle models of atomic phenomena.

Bohm's program is, broadly speaking, a "hidden-variable" reconstruction of the quantum theory. There are two central philosophical issues involved in such reconstruction, both of them intimately related to the central topics of our earlier chapters. One of these issues is the question of whether a theory can or should be modified in the absence of hard empirical data contradicting it. More specifically: does a physicist have the methodological right to tamper with the Heisenberg uncertainty relations and introduce hypotheses conflicting with them when no experimental data are available to show that they require modification?

The second issue centers on a question constantly lurking in the background of our discussions of scope, anomalies, and explanation: what are the limits of explanation within a theory and how are they to be identified? Can a theory be modified to overcome apparent anomalies indefinitely or are there bounds to this process? Does every law have a restriction of scope to some well-specified domain or can some be regarded as truly unlimited in applicability? And so forth.

Regarding the first cluster of questions, we shall see that Professor Bohm regards it as methodologically proper to hypothesize about the possible "breakdown" of the Heisenberg relations in domains not presently accessible to experimental inquiry. The lack of a well-defined anomaly does not deter him from theorizing about the future possibility of our finding such anomalies.

Concerning the second cluster of problems, we shall find Professor Bohm maintaining that all theories and laws of physics are subject to limitations of scope in a very radical way. Within this scope they may hold sway without qualification. Outside it, they must not be assumed applicable without further evidence. The boundaries of explanation, in other words, are for Bohm sharply drawn.

In this chapter, then, it is our intention to survey critically Bohm's program for the reconstruction of quantum theory and his accompanying philosophical justification. Two theses will be defended. First, that Bohm is methodologically correct in proposing hypotheses before a well-defined anomaly is located. For as has already been maintained in Chapter 2, the order of discovery of explanations is independent of the logical requirement that an explanation deal with a natural anomaly.

Secondly, however, it will be argued that Bohm's characterization of scope and the limits of explanation in physical theory involves a fundamental philosophical misconception, a misconception serious enough to call into question the entire "hidden variable" program. In order to defend that charge it will be necessary for us to inquire more closely into the concepts of law, scope, applicability, and anomaly.

THE UNCERTAINTY RELATIONS

Werner Heisenberg begins his elaboration of elementary quantum mechanics and his critique of the wave and particle theories of matter and light with the following remark:

> The starting-point of the critique of the relativity theory was the postulate that there is no signal velocity greater than that of light. In a similar manner, this lower limit to the accuracy with which certain variables can be known simultaneously may be postulated as a law of nature (in the form of the so-called uncertainty

relations) and made the starting-point of the critique which forms the subject matter of the following pages. These uncertainty relations give us that measure of freedom from the limitations of classical concepts which is necessary for a consistent description of atomic processes.[1]

This is basically the thesis against which Professor Bohm sets his antithesis. In particular, it is Bohm's intention to deny to the uncertainty relations the status of "laws of nature" in the special sense that he believes Heisenberg to be employing the term. Bohm's own interpretation is as follows:

The fact that the quantum theory implies that *every* process of measurement will be subject to the same ultimate limitations on its precision led Heisenberg to regard the indeterminacy relationships . . . as being a manifestation of a very fundamental and all-pervasive general principle, which operates throughout the whole of natural law. Thus, rather than consider the indeterminacy relationships primarily as a deduction from the quantum theory in its current form, he postulates these relationships directly as a basic law of nature and assumes instead that all other laws will have to be consistent with these relationships. He is thus effectively supposing that the indeterminacy relationship should have an absolute and final validity, which will continue

---

[1] Werner Heisenberg, *The Physical Principles of the Quantum Theory*, tr. by Carl Eckhardt and Frank C. Hoyt (New York: Dover Publications, Inc., 1930), pp. 3–4.

indefinitely, even if, as seems rather likely, the current form of the quantum theory should eventually have to be corrected, extended, or even changed in a fundamental and revolutionary way.[2]

At first blush, Bohm's words in this passage seem to convey a view of the uncertainty relations as "mere" hypotheses. That is to say, the distinction drawn by Bohm between "a deduction from the quantum theory in its current form" and "a basic law of nature" seems to suggest the kind of dichotomy often cited by philosophers of science as holding between such sentences as 'All crows are black' (a mere contingency) and genuine laws of nature (e.g., Newton's laws of motion). The distinction in this latter case is essentially the difference between a statement which can be falsified by a single counterinstance and a statement (or set of statements) which exhibit a certain tenacity in the face of counterevidence.

But it is evidently not Bohm's intention to make exactly that sort of distinction. For he *does* accept the uncertainty relations as permanent or semipermanent features of our account of the behavior of electrons, photons, etc. On this subject he is quite explicit:

. . . as in the usual interpretation of the quantum theory, we regard the indeterminacy implied by Heisenberg's principle as an objective necessity and not just as a consequence of a simple lack of knowledge on our part concerning some hypothetical 'true' states of the

---

[2] David Bohm, *Causality and Chance in Modern Physics* (New York: Harper and Brothers (Torchbooks), 1957), pp. 83–84.

quantum mechanical variables. Thus, it is not the existence of indetermination and the need for a statistical
theory that distinguishes our point of view from the
usual one. For these features are common to both
points of view. The key difference is that we regard
this particular kind of indeterminacy and the need for
this particular kind of statistical treatment as something
that exists only within the context of the quantum-
mechanical level, so that by broadening the context
we may diminish the indeterminacy below the limits
set by Heisenberg's principle.[3]

The question immediately prompted by this passage is,
of course, whether Bohm is not taking with one hand what
he has given with the other. The answer to this question
depends heavily on the significance of Bohm's conception
of a "context" and what he intends by the phrase "broadening the context." That, as we shall see, is not an easy
matter. Indeed, only when we have examined in some
detail one of Bohm's actual proposals for an alternative
interpretation of quantum mechanics will it be possible
to say in any thorough way what he has in mind. For the
present, a general survey of some of Bohm's remarks on the
notion of a context (or "level") must suffice. This should
shed some light, at least, on the distinction Bohm is attempting to draw between his way of regarding the uncertainty relations and Heisenberg's.

Bohm's doctrine of contexts emerges in the course of his
argument against a position he labels "mechanism." One
of the main failings of this doctrine, he claims, is its failure

---

[3] *Ibid.*, p. 106.

to take contexts into account. As an antidote, Bohm exhorts us to

> . . . recall that no matter how far one goes in the expression of the laws of nature, the results will always depend in an unavoidable way on essentially independent contingencies which exist outside the context under investigation, and which are therefore undergoing chance fluctuations relative to the motions inside the context in question. For this reason, the causal laws applying inside any specified context will evidently not be adequate for the perfect prediction even of what goes on inside this context alone.[4]

The tendency to violate this limitation on laws of nature was, in Bohm's opinion, an important characteristic of —for example—Laplace's claim that a divine predictor, equipped with Newton's laws of motion and gravitation, could predict the entire future course of the universe from knowledge of its present state.

> . . . The conclusion that there is absolutely nothing in the entire universe that does not fit into the general theoretical scheme associated with Newton's laws of motion evidently has implications not necessarily following from the science of mechanics itself, but rather from the *unlimited* extrapolation of this science to all possible sets of conditions and domains of phenomena. Such an extrapolation is evidently then not founded primarily on what is known scientifically. Instead it is

---

[4] *Ibid.,* p. 158.

in a large measure a consequence of a *philosophical* point of view concerning the nature of the world, a point of view which has since that time come to be known as mechanism.[5]

The point is well-taken. As we saw in Chapter 3 the laws of Newtonian mechanics are inherently approximative. To assume that they can be applied without restriction to the universe as a whole is to make an untestable, metaphysical assumption about dynamical closure. In a word, it is a philosophical point of view, not a physical one. There is, however, something odd about Bohm's way of putting the matter. He seems to assume that extrapolation *as such* is to be questioned, not merely extrapolation to systems in which the scope or boundary conditions of the Newtonian laws are not satisfied. We shall return to this critical point a little later on.

Bohm carefully separates two different kinds of mechanistic philosophy. On the one hand there is the deterministic mechanism of Laplace in which all phenomena are presumed to be explicable, in principle, in terms of the motions of point-like bodies. Every event is precisely predictable. This position was gradually undermined by the introduction of fields as relatively autonomous entities in 19th century physics. It is completely repudiated in 20th century quantum mechanics. But in Bohm's view a kind of mechanism has been carried over: indeterministic mechanism. As Bohm sees the Copenhagen interpretation, its mechanistic character far overshadows its differences from the Laplacian philosophy. The tendency to extrap-

---

[5] *Ibid.,* p. 37.

olate beyond the context of applicability of a law and the belief that natural phenomena can be explained in terms of some fundamental set of properties and laws are both retained. Only the insistence upon causal determinism has been sacrificed.

Nevertheless,

> . . . the indeterminacy principle necessitates a renunciation of causality only if we assume that this principle has an absolute and final validity (i.e., without approximation and in every domain that will ever be investigated in physics). On the other hand, if we suppose that this principle applies only as a good approximation and only in some limited domain (which is more or less the one in which the current form of the quantum theory would be applicable), then room is left open for new kinds of causal laws to apply in new domains. . . .

There is good reason to assume the existence of a sub-quantum-mechanical level that is more fundamental than that at which the present quantum theory holds. Within this new level could be operating qualitatively new kinds of laws, leading to those of the current theory as approximations and limiting cases in much the same way that the laws of the atomic domain lead to those of the macroscopic domain. The indeterminacy principle would then apply only in the quantum level, and would have no relevance at all at lower levels. The treatment of the indeterminacy principle as absolute and final can then be criticized as constituting an arbitrary restriction on scientific theories, since it does not follow from the quantum theory as such, but rather from the assumption of the unlimited

validity of certain of its features, an assumption that
can in no way ever be subjected to experimental proof.[6]

In capsule form, then, Bohm's belief is that determinism
can be restored if we reject the mechanistic features of the
usual interpretation. The focal point of his attack on these
mechanistic features remains "the assumption of the unlimited validity" of the uncertainty relations: their extrapolation to contexts (levels) beyond the sphere of their
legitimate employment. The touchstone for the assault is
pointed out once again in this passage:

> Every mechanical law applies only to an isolated system,
> because its behavior depends on boundary conditions
> that are determined in essentially independent systems
> external to the one under consideration. Even if we
> consider the entire universe as a single mechanical sys-
> tem, so that there is no outside, then the same kind of
> problem arises. Thus, when we try to trace the causes
> of what happens at the macroscopic level with greater
> and greater precision, we eventually find dependence
> on the chance fluctuations of the essentially independ-
> ent atomic motions. But these, in turn, depend in part
> on essentially independent chance fluctuations at the
> electronic and nuclear level (as well as on quantum-
> mechanical fluctuations). . . .

> These latter motions in turn depend in part on random
> fluctuations at still deeper levels, connected with the
> structure of the electrons, protons, neutrons, etc. (e.g.,
> mesonic motions and probably even in a level below

---

[6] *Ibid.*, p. 69.

that of the elementary particles). Hence, there is no known case of a causal law that is completely free from dependence on contingencies that are introduced from outside the context treated by the law in question.[7]

The moral? This: the indeterminacy relations, as mechanical principles, are no more free of approximate character than the causal laws of earlier mechanics. They apply in a context or level—the quantum mechanical level—and carry no warranty for use outside that context.

With Bohm's criticisms of what he regards as the prevailing interpretation of quantum mechanics in mind we turn to look more closely at the kind of alternative he wishes to offer. In order to see what sort of alternative he *cannot* offer, however, it will be helpful to consider first a result of von Neumann's bearing on the question of whether the statistical character of quantum mechanics can be removed by an increase in our knowledge of the system under consideration.[8]

Briefly, von Neumann's theorem asserts the following: it is not possible without contradicting the present version of quantum mechanics to add to our present list of observ-

---

[7] *Ibid.,* p. 61.

[8] The discussion of von Neumann's theorem which follows is based on these two papers: James Albertson, "Von Neumann's Hidden-Parameter Proof," *American Journal of Physics,* Vol. 29 (1961), pp. 478–84. I. I. Zinnes, "Hidden Variables in Quantum Mechanics," *American Journal of Physics,* Vol. 26 (1958), pp. 1–4. Von Neumann's original proof is given in his *Mathematische Grundlagen der Quantenmechanik* (Berlin: Springer, 1932; New York: Dover Publications, 1943), see especially pp. 157–173.

ables a new set of data which will enable us to predict the outcomes of observations on individual systems with probability of one.[9] Thus, it is not possible to introduce new observables into the theory in order to eliminate its statistical character. Quantum mechanics is *essentially* a statistical theory: it is not, like classical physical theories employing probabilities, a statistical theory merely as a result of our ignorance of in-principle measurable initial conditions. It cannot seriously be regarded as a theory whose probabilities are all subjective probabilities.

Von Neumann's theorem does not show, of course, that

---

[9] The "new data" are assumed to be data about "hidden parameters" or previously unknown factors within the system representable in the quantum mechanical formalism by Hermitian operators. The general conclusion of von Neumann's argument, incidentally, is developed from two other propositions in the following manner: First, it can be shown that all ensembles of systems whose observables are represented by Hermitian operators exhibit dispersion in the values of the observables. (There are mathematical exceptions to this but they have no physical significance.) Second, it can be shown that all such ensembles are homogeneous, i.e., that the expectation value of any particular observation is the same in any subensemble as it is in the whole ensemble. The main theorem follows from these two propositions. For suppose that a new set of observables is introduced which renders our predictions deterministic. Then it will have done so by dividing the original ensemble into subensembles such that (a) the expectation value of the observable we are now able to predict deterministically is different in the subensemble than in the whole ensemble, and also such that (b) the subensembles are dispersion-free. This contradicts both of the propositions already established.

one cannot construct a new version of mechanics in which the uncertainty relations fail to hold. Many critics have misunderstood this point. It should be obvious, however, that no theorem proved on the basis of the present quantum theory could actually establish that quantum mechanics as presently structured is the only possible mechanical theory. If empirical evidence warrants, the quantum theory may have to be given up. And in that case a new theory— sans uncertainty relations—may supersede it. *So long as no move is made to abandon the present elementary quantum theory, however—so long as its postulates are retained intact—the uncertainty relations cannot be eliminated.* And this holds, says von Neumann, even if we modify the theory in the direction of adding new variables to the state descriptions of electrons and the like. Like Newtonian mechanics, the elementary quantum theory is so structured logically as to be vulnerable only *in toto,* not piecemeal.

Returning to the Bohm program, the significance of von Neumann's theorem is that it rules out the possibility of evading the uncertainty relations in the domain where they presently apply. Bohm is not attempting to do this, of course. Von Neumann's theorem does not, however, rule out the following kind of revision of quantum mechanics: if we assume some sort of sub-quantum level entities and interactions it may be possible to represent the statistical behavior of quantum-level entities. This is the core of Bohm's program.

A classical analogy may help to make the matter clearer. For this purpose Bohm himself refers to the explanation of Brownian motion. This motion, exhibited in the darting behavior of particles in hot smoke, is explained classically on the assumption of random behavior of molecules of the hot gas bumping into the smoke particles. On this basis

statistical predictions about the behavior of the smoke
particle can be worked out.

Of course there is a radical difference between this
classical case and Bohm's program for quantum mechanics.
For in the Brownian motion analysis it was always assumed
to be possible in principle to determine all of the relevant
initial conditions and predict deterministically the behavior of the smoke particle. The probabilities introduced
were thought to be purely a function of our ignorance.
In the quantum mechanical case, on the other hand, neither
Bohm nor anyone else can reasonably propose an interpretation which would allow the statistical features of the
present theory to be eradicated. This puts severe boundary
limitations on the kinds of proposals that can actually be
made.

Bohm's first serious attempt at a reinterpretation of
quantum mechanics in line with his philosophical program came in a pair of papers published in the *Physical
Review* of 1952.[10] We shall give a sketch of some of the
more significant features of that reinterpretation. It should
be borne in mind that Bohm does not currently accept the
results of the 1952 paper as showing more than the possibility of a reinterpretation.

The reinterpretation begins from the premise that an
electron is actually a combination of two distinct features:
a small particle and an associated force field $U(x)$ which
depends upon the position of the particle (here represented by 'x'). Bohm calls the field 'the ψ-field' and we

---

[10] David Bohm, "A Suggested Interpretation of the Quantum
Theory in Terms of 'Hidden' Variables" (in two parts), *Physical Review*, Vol. 85 (1952), pp. 166–93.

shall use both designations. In a sense, what Bohm wishes
to do is to separate the known particle properties of electrons, photons and the like from their wave properties.
The latter, in effect, reduce to being the effects of the force
field U on the particle itself. The method of doing this
mathematically is discussed in Appendix A. Here we need
only remark that Bohm is able to give the field U an appropriate definition in terms of the mathematical formalism of
the quantum theory. Expressions in which the quantum
theory describes the joint wave-particle properties of the
electron are neatly separated into pairs of expressions in
which the behavior of (a) Bohm's particle and (b) Bohm's
U-field are described. An analogy will perhaps be useful
here.

Suppose a biologist is seeking to describe an organism
he has found. But suppose also that the language in which
he is forced to describe it always links together structure
and function; in other words, no part of the anatomy of
the organism can be mentioned without implicitly men-
tioning what the structure does. Instead of 'heart' the biologist must say 'pumping blood,' instead of 'stomach' he
must say 'digesting food,' and so on. A biologist in this
situation is—from Bohm's point of view—very much like
the ordinary quantum theorist. The former cannot separate
structure and function, the latter cannot separate the wave-
and particle-aspects of the electron.

What Bohm does, then, is very much like what the
biologist might do merely by coining a series of anatomical
terms: 'heart,' 'stomach,' 'liver' etc. 'Heart' now designates
the structure, 'pumping blood' the function. In much the
same fashion, Bohm arrives at descriptions of positions and
velocities of a particle (The Particle Aspect) and a description of the field U which disturbs the particle at various

times (Producing the Wave Aspect). The analogy breaks down at the point where we ask: does the Bohm particle —like the structures described by the biologist—really exist? Since the Bohm particle cannot be seen—even with a microscope—the answer to this is not obvious.

The analogy is also shaky on another count. Bohm really does not succeed in *completely* divorcing the wave and particle features. The character of the field U, it must be remembered, depends on the position of the particle x. The field exerts varying forces on the particle depending on the particle's location.

Bohm's electron particle is *not* subject to the Heisenberg uncertainty relations. It can be shown that if the position and velocity of the Bohm particle are known at some time t_1 (and these can, in principle, be established) we can unerringly infer its position and velocity at a later time t_2. Predictive determinism is restored. But, as Bohm quickly points out, this does not have a great deal of practical significance. *For whenever we seek to measure the positions or velocities of the particle the introduction of the measuring instrument disturbs the field* U *in ways it is not mathematically possible to compute.* The force field U becomes, says Bohm, "uncontrollable." Hence we cannot hope to predict the future position and velocity at time t_2 in practice. What we can do is to calculate a distribution of probabilities for the particle's showing up in various places at t_2. And that, of course, is precisely what the ordinary quantum theory would have told us in the first place. There is, however, an important difference. Bohm sums it up this way:

> In the usual interpretation . . . the need for a probability description is regarded as inherent in the very

structure of matter . . . whereas in our interpretation, it arises . . . because from one measurement to the next, we cannot in practice predict or control the precise location of a particle, as a result of corresponding unpredictable and uncontrollable disturbances introduced by the measuring apparatus. Thus, in our interpretation, the use of a statistical ensemble is (as in the case of classical statistical mechanics) only a practical necessity, and not a reflection of an inherent limitation on the precision with which it is correct for us to conceive of the variables defining the state of the system.[11]

Bohm is claiming here that the probabilities of the quantum theory are subjective probabilities: but only with regard to their assignment to his special particle. He does not, so far as I can see, attempt to extend this to the quantum-mechanical electron itself.

But in order even to make good on the assertion that the particle's future positions are unknown to us simply because of ignorance Bohm must specify how, in principle, we *could* measure the initial conditions for the particle. To say merely that these initial conditions are in principle determinable without saying how they are to be determined goes no further than to express misgivings about the truth of the quantum theory. As long as one does not have to say how the measurement is to be carried out, it can always be asserted that the quantum-mechanical electron itself is completely predictable—if only we had enough information. The requirement that a measuring process be specified is crucial in these contexts.

---

[11] *Ibid.*, p. 171.

To meet this need, Bohm supposes that the quantum theory in its current form breaks down when we pass below the level of distances of the order of 10^{-13} cm. or times of the order of $10^{-13}/c$ secs. Within these narrow confines he assumes the existence of a third type of potential affecting the particle—actually, a cluster of small inhomogeneities in the psi-field.

Now the specific difficulty encountered when we try to measure the initial conditions of the particle is that the measurement process perturbs the psi-field uncontrollably. This occurs because

> . . . the potential energy of interaction between electron and apparatus, $V(s, y)$ plays two roles. For it not only introduces a direct interaction between the two particles . . . but it introduces an indirect interaction between these particles, because this potential also appears in the equation governing the ψ-field.[12]

With the assumption of inhomogeneities in the quantum field it becomes at least conceivable that the fluctuations introduced into the psi-field by the measurement process can be neutralized. If this can be done—and on Bohm's interpretation it is at least conceivable—then precise predictions about the particle's future behavior appear to be possible.

CRITICISMS OF BOHM'S INTERPRETATION

Bohm's reinterpretation has not gone unchallenged. In fact, as has already been noted, Bohm himself no longer

---

[12] *Ibid.*, p. 185.

accepts all of the ideas advanced in 1952. Heisenberg has summed up some of these criticisms of the Bohm interpretation. Of particular interest is this one:

> One consequence of this interpretation is, as Pauli has emphasized, that the electrons in the ground states of many atoms should be at rest, not performing any orbital motion around the atomic nucleus. This looks like a contradiction of the experiments, since measurements of the velocity of the electrons in the ground state (for instance, by means of the Compton effect) reveal always a velocity distribution in the ground state, which is—in conformity with the rules of quantum mechanics—given by the square of the wave function in momentum or velocity space. But here Bohm can argue that the measurement can no longer be evaluated by the ordinary laws. He agrees that the normal evaluation of the measurement would indeed lead to a velocity distribution; but when the quantum theory for the measuring equipment is taken into account—especially some strange quantum potentials introduced ad hoc by Bohm—then the statement is admissible that the electrons 'really' always are at rest. In measurements of the position of the particle, Bohm takes the ordinary interpretation of the experiments as correct; in measurements of the velocity he rejects it.[13]

Heisenberg himself rejects Bohm's interpretation in large part because of the asymmetry between velocity (or

---

[13] Werner Heisenberg, *Physics and Philosophy: the Revolution in Modern Science* (New York: Harper and Brothers, 1958), pp. 130–31.

momentum) and position that is involved. In his opinion, it is a fundamental feature of quantum mechanics that it treats momentum and position as closely analogous properties of the electron.

I believe that the criticism being offered against Bohm here may perhaps stem from a misunderstanding encouraged by Bohm's own way of speaking. For the fact is that Bohm's particle is not, as Heisenberg seems to think, the electron of ordinary quantum mechanics. That electron is completely given up as a fiction in the Bohm interpretation (though Bohm never points this out clearly). In its stead we have a particle—which shows up at just those places where electrons do when their positions are measured—and the psi-field. If we choose to refer to Bohm's particle as "the electron" we must be very careful not to confuse it with the electron of ordinary quantum mechanics; for although it occupies the same positions it does not have the same momenta. Thus Heisenberg is not being entirely precise when he says that the Bohm interpretation destroys the quantum-mechanical symmetry between position and momentum. It would be better to say that Bohm abandons the ordinary electron and substitutes a special particle and field for it. When the matter is viewed this way the apparent paradox in an electron's remaining at rest in ground state disappears straightaway.

A second objection—and one taken very seriously by Bohm himself—concerns the generalizability of the treatment of the one-body problem given in the previous section. For it develops that when a generalization to many bodies is attempted, the psi-field becomes a field not of ordinary three-dimensional space but of the multi-dimensional configuration space in which each mechanical coordinate of each body is represented by an axis. The elimination of this problem requires the resolution of certain mathematical difficulties which have so far resisted treatment. Evidently this was one of the factors prompting Bohm to abandon the 1952 interpretation.

Yet another argument against Bohm's approach is the Occamistic criticism that Bohm erects an "ideological superstructure" on quantum mechanics, or graces it with "metaphysical asides." This line of attack is based on the fact, emphasized by Heisenberg, that if we disregard the possibility of modifications of the quantum theory over small distances then "Bohm's language . . . says nothing about physics that is different from what the Copenhagen interpretation says." [14] Accordingly, Bohm's reference to "in-principle" measurable variables and the like are dismissed as superfluous metaphysics.

On Bohm's behalf it must be said this objection hardly carries much force. It is quite true, of course, that quantum mechanics construed à la Bohm has no measurable consequences which quantum mechanics construed à la Heisenberg does not have. But this does not imply that either of the two is talking metaphysics. To the contrary. Let us put it this way: in terms of Popper's demarcation criterion[15] (falsifiability) Bohm's interpretation and the Copenhagen interpretation are equally metaphysical or non-metaphysical, as the case may be. For they are both vulnerable to precisely the same set of empirical findings. Every statement formulated in the Bohm interpretation— including descriptions of the behavior of an electron (or

---

[14] *Ibid.*, pp. 132–33.

[15] Cf. Karl Popper, *The Logic of Scientific Discovery, op. cit.*, pp. 40–42.

"Bohm particle") when it is not being observed—will be at least indirectly falsifiable: just those empirical findings which would prompt a Copenhagen proponent to abandon quantum mechanics in its present form will do the trick. Consequently, Bohm's interpretation does not amount to a metaphysical accretion (in Popper's sense of 'metaphysical') unless we are prepared to regard almost any version of quantum mechanics as metaphysical to an equal extent.

But if Bohm's proposal is not obviously vulnerable to a charge of "metaphysics" it nevertheless is open to objections on other counts. For there are at least two points at which the 1952 interpretation violates rather basic restrictions on the employment of physical concepts.

First consider the status of the psi-field associated with the Bohm particle. As Bohm himself points out, this field is neither emitted nor absorbed. Yet it is assumed to induce changes in the trajectory of the particle with which it is associated. How can this possibly be the case? Obviously, Bohm does not wish to have the particle itself emitting the field. That is ruled out immediately on the assumption that quantum mechanics in its present form is correct. But then we are left with the puzzle: how can the field be construed as always associated with the particle yet capable of acting on that very same particle? The picture which springs directly to mind here is the ancient theory of projectile motion discussed by Aristotle in which air was thought to course around the moving projectile and drive it on from behind. Bohm's psi-field evidently operates in somewhat the same fashion.

All of this is a symptom of the following fact: the properties of Bohm's psi-field cannot be determined by direct measurement. All that we can do is to infer its properties from measurements of the position and momentum of the ordinary quantum mechanical electron. On this ground, if no other, Bohm's contention that the psi-field is objectively real would have to be rejected. A force field cannot legitimately be said to be objectively real if it cannot be shown by observation or direct measurement to affect the behavior of test bodies in determinate ways. Another way of stating the same point is this: assertions to the effect that a field exists in a certain region must always be *directly* testable even though the test is never carried out.

A final difficulty in the 1952 interpretation concerns the question of whether Bohm has succeeded in showing that, in principle, the behavior of his particle can be predicted with certainty. It will be recalled that his argument on this score depended on the postulation of inhomogeneities in the psi-field which could offset the effects of the measuring apparatus. The fact is that this program for measuring the position of the particle without rendering its future behavior unpredictable will not work. The reason is easy to see. In order to be assured that what we are measuring is actually the position of the particle and in order also to know that the particle will not be uncontrollably perturbed by the measurement, we must be able to determine that the intensity and dimensions of the inhomogeneities are not known to be the product of any configuration of bodies, etc. The only way to determine their properties is by empirical investigation. But this means measuring the deviations of a test particle in the psi-field. In other words, Bohm's proposal for the measurement of a particle's position without significantly disturbing it presupposes that we already know how to carry out such measurements or their equivalents. Thus, even if it should be the case that quantum mechanics as we now know it breaks down at Bohm's "sub-quantum" level, it still will not be possible to predict the future behavior of a Bohm particle exactly.

CONTEXTS AND LEVELS

The 1952 Bohm interpretation and arguments offered against it can aid us in becoming clear about the nature of the disagreement between Bohm and the Copenhagen proponents. Especially important is the illumination it throws on Bohm's use of the terms 'context' and 'level,' both in the formulation of his own general program and in his critique of the uncertainty relations as usually understood. One of the things that this terminology refers to, it appears, is the distinction embodied in references to the "quantum level" and the "sub-quantum level" in the formulation of the 1952 interpretation.

Let us recall here Bohm's critique of what he termed the "mechanistic" strain in the Copenhagen interpretation. By this he meant primarily the extrapolation from contexts within which it is not known to apply. He condemns this as a "philosophical" addition to the theory proper; the extrapolation is, for Bohm, wholly illicit anywhere, but particularly objectionable in quantum mechanics.

What he is condemning as "extrapolation" can now be defined a little more precisely. In Bohm's opinion, the uncertainty relations need not be assumed to hold exactly once we begin to deal with processes occurring over short distances and during very brief intervals of time (or involving very high-energy interactions). For Bohm, the alleged applicability of the uncertainty relations to such processes is not a consequence of quantum mechanics or of any experimental argument but a merely "philosophical" assumption.

The trouble with Bohm's way of looking at the matter —and indeed with his entire program—is that he fails to distinguish between several very different senses in which a scientific theory or law is said to "apply" to a subject matter. Because of this, his use of the term 'context' is shot through with a dangerous ambiguity. When all of this is carefully sorted out, we find that his attack on the so-called Copenhagen interpretation rests ultimately on a confusion about the notion of a context.

One important sense in which a scientific law can be said to "apply" or "be applied" is the sense in which the law denotes explicitly a certain class of entities to which it is to be applied. In this sense, Kepler's laws apply to planets; Newton's laws of motion apply to pairs of bodies generally; Snell's law applies to beams of light passing from one medium to another; and so forth. Here, the concept of 'context' stands for something quite definite: a class of objects or entities treated in the statement of the law as subjects of which something is being predicated. Let us call this kind of context the *'extension'* of the law to distinguish it carefully from a second kind of context to be considered below.

The decisive fact about the extension of a scientific law is that it usually will include entities or collections of entities *of which* the law is not literally true. Thus, for example, Kepler's laws are known to hold only approximately for the planets of the solar system. Snell's law fails to hold in the case of such materials as Iceland spar and in other examples of double refraction. Because of this we need to distinguish further between two sorts of extension of a scientific law: the extension for which the law is asserted to be true and the extension for which the law is asserted only to be an approximation. Typically, the first sort of extension will not be found in the physical world but will rather be an idealization or abstraction. In this sense, Kepler's laws apply only to planets which are part of an isolated system of two bodies.

As we found in Chapter 3, there is associated with the applicability of any scientific law a list of what might be termed "boundary conditions." These conditions, when fully articulated, specify the circumstances under which the laws are said to apply. Yet they are very different from the kind of context we called 'extension' a little earlier. In effect, the second kind of context here cited consists of the property or set of properties which the entity or system of entities named in the extension must have if the law is to apply truly. And by 'apply truly' here we mean 'apply in such a way that the law will be construed as false if predictions properly generated from it fail to hold in the subject matter.'

The list of boundary conditions which accompany a given law were referred to earlier as the *scope* of the law. This concept enables us to differentiate more sharply between the two types of extension a law will have. That part of its extension or class of applications in which all scope conditions are satisfied we can call the *scope-extension* of the law. The total class of applications, including both approximative and exact uses of the law, can be called the *total-extension*.

In these terms, the various senses in which a law can be said not to apply to a subject matter can be put in four classifications. (1) It may be that neither the scope conditions are satisfied in the subject matter nor the objects of the

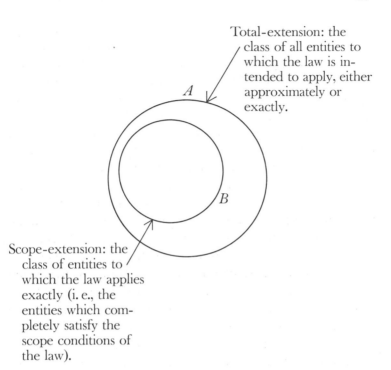

Total-extension: the class of all entities to which the law is intended to apply, either approximately or exactly.

A

B

Scope-extension: the class of entities to which the law applies exactly (i. e., the entities which completely satisfy the scope conditions of the law).

total-extension contained in it. In this sense, the law fails to apply because it is irrelevant. The objects lie outside the circle *A* on the diagram above. (Example: attempting to apply Snell's law to the analysis of a planetary orbit.) (2) It may be that the scope conditions are met but that the subject matter does not include objects of the total-extension. Again we are outside circle *A*. This too is a rather uninteresting case. (Example: attempting to apply Newton's inverse-square law of gravitation to two massless points of a closed, field-free region of empty space.) (3) A more interesting kind of case arises when the subject mat-

ter contains objects of the total-extension but the scope
conditions are not met. In this circumstance, we *may* say
that the law applies approximately if we can satisfy ourselves that fulfillment of the scope conditions would make
the law apply truly to the subject matter at hand. But we
may also decide that either (a) the law would not apply
truly upon fulfillment of the scope conditions—implying
the falsity of the law—or (b) we do not know how to fulfill
the scope conditions. *In case (a) the observations implying
the falsity of the law will constitute observations of anomalous states of affairs.* In case (b) we should ordinarily have
to regard the law as failing to apply in the sense of being
useless for predictive and explanatory purposes. This could
be a permanent situation or only temporary depending
upon the possibilities of our ever being able to fulfill the
scope conditions. On the diagram above, cases (3a–3b) correspond to points in *A* but outside *B*. Finally, (4) it may be
that the scope conditions are met and the objects of the
extension are contained in the subject matter but that the
law simply does not hold true for the subject matter. This
is the case in which the law is said to be unconditionally
false. The only alternatives are to abandon it completely
or to modify its scope conditions in some such way as will
provide for an explanation of the anomaly.

   Rather than pursue this analysis further, let us ask:
which sense of 'context' is Bohm employing in his criticism
of the Copenhagen interpretation? Scope-extension or total-
extension? Actually, both. But with a rather unusual twist.
*For he assumes that context in the sense of scope-extension
always has a restriction to entities or processes of a certain
magnitude.* There is always, for Bohm, a kind of threshold
below which the law is not necessarily supposed to apply.
Look again at Bohm's own words:

. . . every mechanical law applies only to an isolated system, because its behavior depends on boundary conditions that are determined in essentially independent systems external to the one under consideration. Even if we consider the entire universe as a single mechanical system, so that there is no outside, then the same kind of problem arises. Thus, when we try to trace the causes of what happens at the macroscopic level with greater and greater precision, we eventually find dependence on the chance fluctuations of the essentially independent atomic motions. But these, in turn, depend in part on essentially independent chance fluctuations at the electronic and nuclear level (as well as on quantum-mechanical fluctuations . . .). These latter motions in turn depend in part on random fluctuations at still deeper levels. . . . Hence, there is no known case of a causal law that is completely free from dependence on contingencies that are introduced from outside the context treated by the law in question.[16]

Bohm's vacillation in this passage between context as scope-extension ("boundary conditions") and context as an indication of the approximate size of the entities dealt with infects the whole conception of a "context" with ambiguity. Certainly it is correct to say that all (or at least most) mechanical laws apply only under certain boundary conditions. But it is not correct to imply that those boundary conditions are *essentially* connected with the size of the objects included in the total-extension of the law. Any such connection between the size of the entities dealt with and

---

[16] Bohm, *Causality and Chance, op. cit.,* p. 61.

the general boundary conditions which the entities must
satisfy would be a peculiar feature of particular mechanical
laws, not a defining characteristic of mechanical laws in
general.

An example may help to make this more evident. Newton's law of gravitation is applicable only in certain limited
contexts; that is to say, it applies only under certain boundary conditions. The objects to which the law is applied in
drawing inferences must, for example, constitute a closed
two-body system. But none of these boundary conditions
mentions the *size* of the bodies to which the law can be
applied. As initially conceived, Newton's "bodies" were
nothing more than punctiform masses or objects of arbitrary size whose distribution of mass could be construed as
occurring at a single point. Only with the collapse of Newtonian mechanics in the 19th and 20th centuries does size
become a factor at all.

The key point is this: the law as conceived prior to the
20th century *was* subject to certain limitations (boundary
conditions) governing its applicability. But these limitations had nothing whatever to do with size of the objects
dealt with. It is only when the Newtonian law is recognized
to be false that the factor of size is brought into play.

The analogy with Newton's law has misled some into
believing that Bohm *must* be taking the statements of the
uncertainty relations to be false when he proposes to regard
them as "not applying" in the sub-quantum mechanical
domain. This, of course, would contradict Bohm's avowal
that he takes the uncertainty relations to be objectively
necessary just to the degree that they are so regarded in the
current quantum theory. But he is not bound to regard
the uncertainty principles as false statements in order to
say that they do not apply. He has, I believe, three options

available: (a) he may assume that the scope conditions of the uncertainty relations are not met in the small-scale phenomena; (b) he may assume that the total-extension of the uncertainty relations does not include the objects of the so-called sub-quantum domain; or (c) he may say that the scope conditions are met, the total-extension includes the entities dealt with, but the uncertainty relations do not obtain. These three options derive directly from our analysis of the ways in which a law or principle can be said not to apply in a context.

Which of these three options does Bohm actually pick? The ambiguity of his language makes it difficult to say. For when he talks *about* his program he seems to envision new kinds of entities and properties in the sub-quantum level such that the uncertainty relations "don't apply" in either sense (a) or sense (b).[17] In the actual details of the 1952

---

[17] Consider, for example, the following passage (*ibid.,* p. 94): "Let us begin with a discussion of the indeterminacy principle. We recall that in the proof of this principle, it was essential to use three properties: namely, the quantization of energy and momentum in all interactions, the existence of wave-like and particle-like aspects of these quanta, and the unpredictable and uncontrollable character of certain features of the individual quantum process. It is certainly true that these properties follow from the current general form of the quantum theory. But the question we raised . . . was precisely that of whether or not there exists a deeper sub-quantum mechanical level of continuous and causally determined motion, which could lead to the laws of quantum mechanics as an approximation holding at the atomic level. For if such a sub-quantum mechanical level exists, then as we have seen, *the basic assumptions cited above,* which are necessary for the validity of the indeter-

interpretation, however, he does find it necessary to assume the falsity of the uncertainty relations in order to describe the process of measuring a particle's position without disturbing it uncontrollably. The exploitation of this ambiguity is of central importance to the enterprise he is about. Let us try to say why.

If either option (a) or (b) were to be selected, the kind of mild theoretical revision proposed by Bohm in 1952, a revision in which quantum mechanics in its present form is retained almost intact, would be a reasonable possibility. For in this case the uncertainty relations would simply fail to apply in the new domain in the way that Snell's law fails to apply in double refraction. When we examine options (a) and (b) in detail, however, we find that they are simply not tenable. The boundary conditions under which the uncertainty relations apply and the kinds of entities to which they apply are so broad that there is no way to describe an intelligible set of physical objects to which they are not applicable in the specified sense. As Bohm himself acknowledges, the thrust of the experimental findings which led to the establishment of these relations was such as to support the contention that *any* form of matter or energy would be subject to them in the measurement situation. To suppose, therefore, that the sub-quantum mechanical level contains entities to which they do not apply in either sense (a) or (b) is to entertain the possibility of

---

minacy principle, *would not hold at this lower level."* (My italics.) It is far from clear what Bohm means when he says in this paragraph that the three conditions "follow from the current general form of the quantum theory." That they obtain in any given situation is *not,* as Bohm seems to suggest, something which can be learned without reference to the facts.

very strange entities indeed. It amounts virtually to a willingness to stop doing physics.

Whenever a critic points out this difficulty, however, Bohm can argue—and perhaps justifiably in some cases— that the critic is excessively conservative. Here, option (c) is brought to the fore and the critic is reminded that, after all, the uncertainty principles may very well turn out false in the small-scale domain. Any reasonable critic will be willing to grant the point: quantum mechanics *is* a physical theory and it may be overthrown by new findings.

The problem for the Bohm program, however, is that although this tactical move has a certain rhetorical value it results in an embarrassing situation. For if option (c) is actually selected, the modest proposals of Bohm will surely be inadequate: no amount of tinkering with the equations of quantum mechanics will suffice if the uncertainty principles are found to be false. A complete revolution—comparable to the revolution of quantum mechanics itself— would be required in our way of looking at particle physics.

Thus Bohm vacillates between the three options open to him. To settle on one or another would completely undermine his program. For if (c) is selected a burden of experimental proof goes with it. It must eventually be shown that genuine anomalies exist with respect to the uncertainty relations. Bohm's predictions, like those of Einstein in 1905, must eventually be borne out by observations conclusively disconfirming the quantum theory's most basic laws. And a scientific revolution of major proportions—not just a reworking of present theories—would ensue if that kind of disconfirmation were forthcoming. On the other hand, choosing (a) or (b) requires that the impossible be achieved: namely, that entities be described to which the uncertainty relations do not have a *prima facie* claim to apply.

In concluding our exposition and critique of Bohm's program let us last of all return to Bohm's treatment of the philosophical position he calls mechanism. As we have said before, the key element of mechanism for Bohm is its tendency to extrapolate beyond the context in which a scientific law or principle actually applies.

It should be evident by now that 'extrapolate' as Bohm uses the term shares in the ambiguity of 'context' and 'applies.' When employed to refer to attempts to extend laws into domains where either their scope conditions are not met or the entities of the domain are not part of their total-extension the term 'extrapolate' has much to recommend it. For in such cases, e.g., Laplace's demon, the use of the law (or laws) turns out to be a misuse.

But it is clearly a mistake to regard applications of the uncertainty relations to bodies of sub-quantum dimension or applications of Newton's laws to closed systems of bodies (however small) as illicit extrapolations. *It is part of the meanings of such principles and laws that they are supposed to apply in such circumstances.* By 'apply' here we emphatically do *not* mean 'turn out to be true.' There is no assumption of "final and ultimate validity of the theory," no presumption that the law or principle will never be found to be false. Instead, we mean only to assert the *prima facie* obligation of the law or principle to provide us with grounds for inferences about the subject matter.

In spreading the blanket condemnation "extrapolation" over a few sins, Bohm succeeds in covering up a number of virtues. In pointing out correctly that mechanical laws are restricted in their applicability he simultaneously obscures the actual generality of those laws. The cost of a too-rich philosophical language is nowhere more evident.

We must reiterate, however, the stand taken at the out-set regarding Bohm's—or indeed anyone's—methodological right to hypothesize in the absence of a well-defined anomaly in the quantum theory. Dogmatic insistence that a contradiction between theory and observation must be produced *before* tentative explanations are put forward is wholly unwarranted by the logic of explanation. Without free speculation such contradictions may never be found. As Einstein clearly saw in 1905, it is sometimes necessary to adopt an "heuristic standpoint"—to offer explanations before their need is logically demonstrable. Our quarrel here is not with Bohm's right to speculate and hypothesize. It is only with the *kind* of speculation he has so far chosen to offer and, more importantly, with his philosophical justification of its character.

6

Anomalies and Statistical
Explanation in Meson Theory

It is a continuing puzzle—far from settled in all details today—how the nucleus of an atom remains in one piece. An elementary consideration locates the problem quite well: a helium nucleus consists of two protons bearing one unit of positive electrical charge each and two neutrons bearing no charge but having about the same mass, etc., as the protons. As any high school physics student knows, like electrical charges repel one another. Why then doesn't the nucleus of helium fly apart under the influence of the protonic charges within it? Why doesn't the universe fall into a completely chaotic state of radioactive confusion?

The story of how physicists have attacked this problem in the 20th century is surely one of the most engrossing chapters in the history of science. Much of it, as we said, remains to be written. That part of it concerned with the discovery of the meson, however, can already be seen to constitute the central thread. In this chapter we shall examine the early history of the meson theory. The aim is to shed light on the ways in which the development of knowledge in this basic branch of modern physics has occurred. Many of the concepts which have occupied our attention so far—anomalies, statistical laws, scope, approximation, the structure of theories—are involved in the telling of the story. Our aim, however, is historical rather than logical: We seek to examine the dynamic process of scientific investigation in its natural setting. Philosophical morals, therefore, will be kept to the barest minimum possible. Yet it goes without saying that they are there to be drawn and the perceptive reader will find it easy to do so. *Seen in proper perspective the history of the discovery of the meson is the record of ever-deepening penetration into the nature of matter through the exploration, definition, and explanation of anomalies.* Remarkably, a single theoretical formalism—quantum mechanics—provides the conceptual background for the entire story. Repeatedly challenged, constantly in danger of overthrow, the quantum theory triumphs in the work of Yukawa in a fashion reminiscent of Leverrier's Newtonian victory over Uranus.

BEFORE YUKAWA

Like Hertz's initial problem about the photoelectric effect, the problem about nuclear binding forces set out above is a gross one. Indeed, the level here is even more superficial. For we already know of other forces which might account for the binding energy of the nucleons (protons and neutrons) besides electrical attractions. Newtonian gravitation, for example, is an obvious candidate. All massy particles attract one another with a force inversely proportional—or nearly so—to the square of their distance from one another, and this attractive force is independent of the attractive and repulsive forces induced by electrical charge.

Unfortunately, the gravitational force proves inadequate to the task of explaining nuclear binding. This is obvious from macrophysical considerations. The leaves of an electroscope, for example, would never part if the Coulomb forces generated by the like charges on both leaves were smaller than the gravitational forces between the particles making up the leaves. Compared to the Coulomb force, gravitation is extremely minute.

Similar remarks hold for other forces which might be suggested. In fact, by 1934, when the idea of the meson was first formulated, empirical investigations had revealed that nuclear binding forces are so strong as to dwarf the Coulomb force by contrast. The problem of ascertaining their nature seemed to be slightly simplified by the discovery of the neutron in 1932 by Joliot, Curie, and Chadwick.[1] By considering the binding force between a neutron and a proton (rather than two protons) the matter of charge could be completely disregarded. Werner Heisenberg describes the state of knowledge at the time in this fashion:

> . . . experience concerning the deflection of protons by protons proves that forces of attraction are acting between particles of the same kind—in other words, not only between protons but between neutrons, too— which forces of attraction are approximately equal to those acting between protons and neutrons. In the case of two protons the situation is more complicated, because the electric force of repulsion is superimposed on the nuclear force of attraction. But when dealing with

---

[1] See I. Curie and F. Joliot, *Comptes Rendus,* Vol. 194 (1932), pp. 273 and 876; J. Chadwick, *Nature,* Vol. 129 (1932), p. 312.

very short distances, the force of repulsion is much weaker than the nuclear force, so that in this case, practically, only the latter is operative. However, due to its long range, the electric force continues to be perceptible long after the nuclear force has ceased to be operative. If we draw a diagram of the potential energy of a proton at various distances from another proton, it will look more or less like the one [in Figure A]. In fact, up to a distance of the order of 5×10^{-13} cm., the picture is practically identical with the one shown in [Figure B]. From that point on, however, the potential energy does not approach 0 asymptotically, but passes through 0, rises to positive magnitudes, and only then does it drop asymptotically toward 0.[2]

Besides the general problem described by Heisenberg, anomalies also were plaguing at least two other areas of nuclear physics during the period from 1930 to 1934. The two areas in question were (a) the observational study of cosmic rays by Geiger-counter coincidence and Wilson cloud chamber techniques, and (b) the investigation—primarily theoretical—of the nature of the beta-decay of radioactive nuclei. Let us describe these in turn.

(A) *The cosmic rays.* In the early 1930's two general hypotheses were current concerning the nature and origin of cosmic radiation. They took the form of answers to this question: is the primary or original component of the cosmic rays striking the upper atmosphere made up of high-

---

[2] Werner Heisenberg, *Nuclear Physics,* tr. by Frank Gaynor and Amethe von Zeppelin (London: Methuen and Co., 1953), pp. 94–95. Figures A and B follow those of Heisenberg.

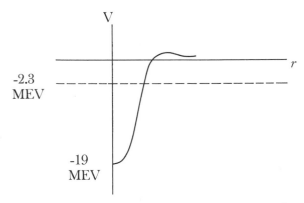

Figure A. Potential of the force between proton and proton

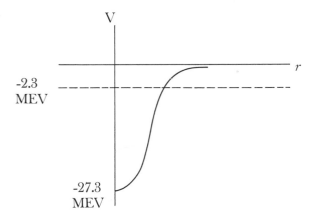

Figure B. Potential of the force between neutron, proton

energy photons (as had initially been supposed by nearly everyone before 1929) or is it made up of charged particles (as W. Bothe and W. Kohlhorster[3] had suggested in that year)? We now know that neither of these hypotheses is strictly correct or strictly incorrect: the cosmic ray primary component consists of a few heavy atomic nuclei stripped of their electrons, a great many protons, some electrons and even a few photons. But this complex constitution was not completely disclosed until about 1961.[4]

In the course of evaluating these two hypotheses on the nature of the primary cosmic rays during the early 1930's a whole host of scientists from various nations undertook a search which carried them to all parts of the world. As a result, much information came to light not only about the primary component but also about that part of the cosmic radiation which is generated *within* the atmosphere as a result of the impingement of the primary rays. For one thing, it was discovered that the primary radiation gives rise to certain "secondaries" which, in turn, are capable of producing "showers" or bursts of radiation within the atmosphere. Furthermore, it was discovered by Carl Anderson in 1932 that many of the particles in the secondary radiation were "positrons"—the positive electrons forecast by P. A. M. Dirac on the basis of the relativistic quantum theory.[5] In fact the secondary radiation seemed to be composed about equally of positive and negative electrons. At

---

[3] For a discussion of their work see Bruno Rossi, *Cosmic Rays* (New York: McGraw-Hill Book Co., 1964), pp. 35–42.

[4] *Ibid.,* p. 167; also V. L. Ginzburg and S. I. Syrovatsky, *Journal of Theoretical Physics* (Supplement), No. 20 (1961), p. 1.

[5] C. D. Anderson, *Physical Review,* Vol. 43 (1933), p. 491.

about this time Bruno Rossi and others succeeded in showing also that another part of the secondary radiation had an amazing capacity for penetrating through the atmosphere and through lead plates.[6] This second "penetrating" component did not generally cause showers to occur and appeared to be almost entirely positive in charge. (*See Figure C.*)

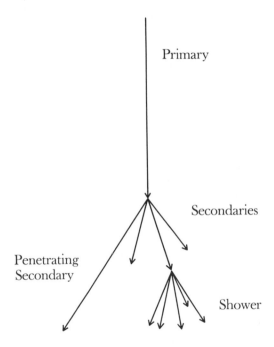

Figure C. Schematic Illustration of the Taxonomy of Cosmic Ray Particles

[6] Rossi, *op. cit.*, pp. 45–46.

Now by 1934 the large-scale inquiry into the nature of the cosmic rays had really begun to bear fruit. It was discovered that variations in the overall intensity of the radiation were noticeable depending on whether the observer was facing the eastern or the western part of the sky. It also seemed to be possible that variation in intensity was correlated with latitude of the observer in some way, though this was far from established. At any rate, these variations lent strong credence to the view that the primary radiation consists of charged particles, attracted perhaps by an envelope of electric charge about the earth (as T. H. Johnson suggested), and deflected in a fairly uniform way by the gravitational field of the earth.

The primary particles would have to have a positive charge because of the direction in which they were deflected by the earth's magnetic field. This meant they were either protons or positrons. The highly penetrating secondary component also had been found to exhibit east-west and latitude effects implying it to be of positive charge. Hence it consisted either of protons or positrons. And, finally, so did about half of the "soft" secondary radiation which produced showers. The question in each case was: proton or positron? Only for the last of these three was the answer unambiguous. Anderson's data on ionization showed clearly that these shower-producing "soft" rays were of electron mass.

In general, the question remained open in the mid-30's so far as the primary radiation was concerned. Advocates of the corpuscular hypothesis let their case rest on the basis of the "east-west" variation and turned their attention to the two secondary components of positive charge. And here it was that an important anomaly arose. Comparative measurements of the energy, ionization, penetration, and ab-

sorption of these two components showed striking differences but, at the same time, striking similarities. Something particularly seemed wrong about the great energy
loss experienced by the "penetrating component" as compared with the much smaller energy loss experienced by
the "shower-producing component." [7]

The options at the time seemed to most investigators to
be the following: (a) The penetrating component might
be taken to consist of protons, rather than positrons. The
difficulty with this was that at least a few of the penetrating
particles seemed to be negatively charged. Moreover,
whether or not one wished to assume the existence of negative protons, the ionization of both positive and negative
penetrating particles looked more like the ionization of

---

[7] T. H. Johnson, *Physical Review*, Vol. 45 (1934), p. 569: "Although the corpuscular radiation is widely distributed in
energy, close analysis shows a band of greater-than-average
intensity in the range of positron or proton energies from
1 to 1.8×10^{10} volts. Independent determinations of energy
and absorption coefficient show disagreement in order of magnitude with the theory of energy loss by ionization, and the
atmospheric range of the . . . radiation is anomalously independent of primary energy. Both processes point to some other
process for dissipation of energy." What brings on this perplexity is that a penetrating particle ionizes a great many
more molecules of gas than a shower-producing particle with
the same initial energy. If we assume that energy loss due to
radiation is the same for both particles then the total energy
lost by the penetrating particle appears to be greater than that
lost by the shower-producing particle. But that is impossible,
for they both start with the same energy and both end up at
rest.

particles of electronic mass than of protonic mass. (b) The
penetrating particles could be assumed to be positrons with
the proviso that the differences between their behavior and
the behavior of the non-penetrating shower-producing rays
was a function of the shower-production mechanism. Since
this mechanism was not at the time perfectly understood
such an assumption had the merit of being promising if
nothing else. As T. H. Johnson put it in 1934, ". . . shower-
production possibly accounts for the absorption anomalies." [8]

As we shall see, the gradual development of understanding of the shower-production process and of the processes
of energy loss through ionization and radiation led in the
end to the elimination of both options (a) and (b) as live
possibilities. The second of them—more plausible from
the start—went down hardest. *For it was based on a fundamental misconception of what needed explaining.* It was
not the absorption behavior of the shower-producing particles which was anomalous—as Johnson and others assumed—it was the absorption behavior of the penetrating
particles! Only with the development of the quantum
theory of shower-production could this be determined.
Without a fully developed theoretical background it would
be impossible to say whether the behavior of the penetrating component was "natural" and the non-penetrating
component "unnatural" or vice versa. Even before the full
elaboration of theory, however, the existence of an anomaly
was assured. Either the penetrating or shower-producing
particles were behaving in ways incompatible with the
obvious explanations.

---

[8] *Ibid.*

Whichever particle was construed to be the peculiar one, the anomaly was established—a natural anomaly, we should note, concerning the behavior of the individual secondary rays. Ordinary considerations of mechanics dictate that all of the secondaries exhibit the same energy loss, given that all are of the same mass and initial energy. One does not even require quantum mechanical reasoning to establish the point; the conservation of energy principle will do. Where quantum theory is needed, however, is in quantitatively determining which set of particles has the "wrong" energy losses. We shall return to that aspect of the story momentarily.

(B) *Beta-decay.* The second area from which illumination was to come was that of the study of so-called beta-decay of atomic nuclei. "The atomic nuclei, which consist only of protons and neutrons, can emit either electrons or positrons by beta-decay." [9] The question is: how? How exactly can a cluster of massive protons and neutrons give rise to the smaller electrons and positrons—particularly in view of the negative charge of the former and the absence of negative charge in the nucleons?

In 1932 and 1933 Heisenberg undertook a detailed analysis of the role of the neutron in the nuclear structure. The prior state of the subject is reviewed briefly in the following paragraph and hints of his *modus operandi* in what was to follow also appear.

Through the researches of Curie and Joliot and their interpretation by Chadwick it has been established that

---

[9] C. F. von Weizsäcker, "Theory of the Meson," in *Cosmic Radiation,* ed. Werner Heisenberg (New York: Dover Publications, 1946), p. 98.

a new fundamental element, the neutron, plays an important role in the structure of the atomic nuclei. This result suggests the assumption that atomic nuclei are constructed of protons and neutrons without the inclusion of electrons. If this assumption is correct it implies an extraordinary simplification for the theory of the atomic nucleus. The fundamental difficulties which one encounters in the theory of β-decay . . . can then be reduced to the question of the ways in which a neutron can decay into a proton and electron and which statistics it satisfies, while the actual structure of the nucleus can be described by considering the forces between protons and neutrons in accordance with the laws of quantum mechanics.[10]

Heisenberg's central hypothesis, therefore, would be that the electron emitted in beta-decay was a by-product of

---

[10] Werner Heisenberg, "Uber den Bau den Atomkerne I," *Zeitschrift für Physik,* Vol. 77 (1932), p. 1: "Durch die Versuche von Curie und Joliot und deren Interpretation durch Chadwick hat es sich herausgestellt, dass im Aufbau der Kerne ein neuer fundamentaler Baustein, das Neutron, eine wichtige Rolle spielt. Dieses Ergebnis legt die Annahme nahe, die Atomkerne seien aus Protonen und Neutronen aufgebaut. Ist diese Annahme richtig, so bedeutet sie eine ausserordentliche Vereinfachung für die Theorie des β-Zerfalls . . . begegnet, lassen sich nämlich dann reduzieren auf die Frage, in welcher Weise ein Neutron in Proton und Elektron zerfallen kann und welcher Statistik es genügt, während der eigentliche Aufbau der Kerne nach den Gesetzen der Quantenmechanik aus den Kraftwirkungen zwischen Protonen und Neutronen beschrieben werden kann."

the alteration of a neutron into a proton. Conservation of charge would then suffice to answer the question we first raised above; viz., how can an electron with negative charge be emitted from the positively and neutrally constituted nucleus? Heisenberg's hypothesis, however, is intended to solve a far deeper problem than this. The superficial question of charge conservation is really unimportant compared to the profounder anomaly which only emerges in the later portions of Heisenberg's report of his research. There he comes to consider the possibility that another general hypothesis might do as well to describe and explain the beta-decay. The alternative is the assumption that the neutron is not itself a fundamental particle but rather a composite formed by the union of an electron and proton. On this view, the electron emitted in beta-decay is not *produced* by the "decay" of the neutron; rather, it is *released* from its position of bondage. Nuclei then would be nothing more than clusters of protons and electrons.

In the nucleus as Heisenberg wishes to depict it, on the other hand, the basic constituents are protons and neutrons construed as distinct states of a single particle. Electrons are produced as concomitants of the change of state and need not actually be assumed to exist *as particles* within the nucleus but can be interpreted as an interaction field binding them together. The opposed hypothesis would describe the hydrogen nucleus as consisting of two protons and one electron. This leaves the ultimate source of the binding energy of the nucleus an open question.

In support of his own hypothesis and against the alternative view Heisenberg argues as follows:

. . . one can first of all mention that *the existence of the neutron already contradicts the laws of quantum*

mechanics in their present form. Not only the initial
hypothetical validity of Fermi statistics for neutrons
but also the failure of the energy relations in Beta-
decay proves *the inapplicability of the earlier quantum
mechanics to the structure of the neutron*. But even if
one ignores these properties of the neutron, the fact
that it is an object with the approximate position
spread $\Delta q \sim e^2/mc^2$ already implies a contradiction of
quantum mechanics if one conceives the neutron as
constructed out of a proton and electron. [My italics.][11]

The problem about the alternative hypothesis, therefore, is
that it leads to a contradiction of basic laws of quantum
physics: a position spread of the indicated magnitude, if
taken to apply to an object which is a compound of proton
and electron, leads to such consequences as that the neutron (proton-electron) must have a mass defect of about 137
mc^2. This is more than 100 times the actual value.

Moreover the alternative hypothesis does not even begin

---

[11] Werner Heisenberg, "Uber den Bau den Atomkerne II,"
Zeitschrift für Physik, Vol. 78 (1932), p. 163: ". . . [man kann]
zunächst anführen, dass schon die Existenz des Neutrons den
Gesetzen der Quantenmechanik in ihrer bisherigen Form
widerspricht. Sowohl die allerdings hypothetische Gültigkeit
der Fermistatistik für Neutronen, wie das Versagen des Ener-
giesatzes beim β-Zerfall beweist die Unanwendbarkeit der
bisherigen Quantenmechanik auf die Struktur des Neutrons.
Aber selbst wenn man von diesen Eigenschaften des Neutrons
absieht, so bedeutet bereits der Umstand, dass das Neutron
ein Gebilde der ungefähren Ausdehnung $\Delta q \sim e^2/mc^2$ ist, einen
Widerspruch zur Quantenmechanik, wenn man das Neutron
als zusammengesetzt aus Elektron und Proton auffasst."

to explain away the original difficulty. This difficulty Heisenberg tersely notes when he speaks of "the failure of the energy relations in β-decay." What he means is that there was every reason to believe in 1932 that the phenomenon of beta-emission constituted a violation of the conservation of energy principle. To put the matter in a somewhat oversimplified way, it appeared that electrons were being emitted from identical nuclei with different velocities—indeed, a continuous spectrum of velocities—while the nuclei all ended up in identical states again. Energy evidently was being dissipated somewhere but no trace of it could be found. Heisenberg's hypothesis at least held out the hope of resolving this difficulty.

It becomes clearer here that Heisenberg's undertaking in attempting to fix the role of the neutron in the nucleus involves more than a mere resolution of superficial difficulties about charge conservation. Other observed characteristics of the neutron and nuclear radiation give rise to perplexities of a more serious nature and it is these perplexities which his account seeks to clear up.

Most of the anomalies here—unlike the absorption anomalies in secondary cosmic radiation—are of the statistical type. They concern the distributions of energies of the emitted beta-rays and the statistical behavior of the newly discovered neutrons. Yet they threaten most directly the (non-statistical) principle of conservation of energy! We see here an illustration of a major point discussed in Chapter 4: statistical theories like quantum mechanics are not vulnerable piecemeal but only in a wholesale way. When the expected distribution of beta-ray energies fails to occur one cannot simply write down the actual distribution in one's notebook and chalk the error up to experience (though that is precisely what an insurance actuary would do with new data about the frequency of death in a population). The underlying theoretical structure from which the predicted distribution was deduced must be carefully re-examined for possible ways of explaining away the difficulties—which is just what Heisenberg was seeking to do.

The bulk of Heisenberg's work is devoted to determining necessary and sufficient conditions (compatible with the quantum theory) for beta-emission from a nucleus. In the end he is forced to concede that no completely adequate criterion has been found. Along the way, however, a number of pertinent and provocative suggestions are made which bear on the problem of the nuclear binding forces. We shall trace out here only enough of Heisenberg's account to indicate the nature of his suggestions and their implications for the later work of Fermi and Yukawa.

An analogy forms the starting point for Heisenberg's development of a formalism to describe the neutron's role in the nucleus. The analogy is that between the (i) so-called "exchange force" between two molecules resulting from their "sharing" a valence electron and (ii) an interaction force between a neutron and proton. The nucleus, then, is being seen as a kind of miniature molecule in which neutrons and protons take on the roles of atoms and, like atoms, share an electron. The electron, however, is not a valence electron. In fact, on account of the shortness of intranuclear distances and on account of the uncertainty relations it is not really meaningful to speak of the electron as being spatially "exchanged" between the two nucleons. It is better to adopt the language of fields and say simply that an exchange force acts between the two nucleons in such a way that they change their state (proton to neutron or vice versa) simultaneously or else emit an electron should one change its state alone. Visual metaphors about the neutron and proton tossing an electron back and forth between them are therefore to be avoided. "It is surely more correct to view the exchange force integral $J(r)$ as a fundamental characteristic of the neutron-proton pair without seeking to reduce it to motions of electrons." [12]

Heisenberg's employment of the molecular analogy to arrive at a mathematical description of interaction forces is a brilliant example of the heuristic employment of analogical reasoning in modern physical theory. By construing the nucleus as a kind of miniature molecule he brings to bear the developed mathematical apparatus required for inferences about nuclear goings-on. Bohr had done something similar many years before in viewing the atom as a miniature planetary system with negatively charged electrons playing the planetary parts. It must be stressed, however, that the use of analogical reasoning is nothing more than heuristic. It is a gross genetic fallacy to assume that because physicists *discover* formal descriptions and theories by means of analogies that the validity or truth of those descriptions is in some way guaranteed by the analogy.

However brilliant Heisenberg's reasoning and however illuminating were his suggestions about the nuclear binding forces, the fact remains that he did not (and perhaps could not) account for the phenomenon of beta-emission by means of his hypothesis alone. The continuous spectrum of velocities of emission remained an anomaly and

---

[12] Heisenberg, "Uber den Bau . . . I," *op. cit.*, p. 12: "Es ist aber wohl richtiger, das Platzwechselintegral $J(r)$ als eine fundamentale Eigenschaft des Paares Neutron und Proton anzusehen, ohne es auf Elektronenbewegungen reduzieren zu wollen."

not even the conditions under which emission should be expected to occur were completely certain. The problem had been brought into focus, though, and great leaps forward could be anticipated in the understanding of the nuclear structure if the problem could be solved.

The problem was solved—at least to a first approximation—in the following year when Enrico Fermi published his "Versuch einer Theorie der Beta-Strahlen I." [13] Combining Heisenberg's hypothesis with Wolfgang Pauli's[14] assumption that beta-emission involves an as yet unobserved neutral particle of small mass (the neutrino), Fermi succeeded in providing a reasonably complete and extremely deep formal treatment of the beta-decay issue.

Fermi's procedure in setting up a formalism begins with the assumption that there are no light particles (electrons, etc.) constituent in the nucleus but that two such particles are emitted or absorbed every time a heavy particle changes its internal state (i.e., changes from proton to neutron or vice versa). More exactly: he assumes that an electron and neutrino are absorbed by the nucleus when a proton becomes a neutron and emitted whenever a neutron becomes a proton. He then identifies three components of the

---

[13] *Zeitschrift für Physik,* Vol. 88 (1934), pp. 161–77.

[14] *Ibid.,* p. 161: "Nach dem Vorschlag von W. Pauli kann man z. B. annehmen, dass beim β-Zerfall nicht nur ein Elektron, sondern auch ein neues Teilchen, das sogenannte "Neutrino" (Masse von der Grössenordnung oder kleiner als die Elektronenmasse; keine elektrische Ladung) emittiert wird." ["Following a proposal of W. Pauli one may for example assume, that in β-decay not only an electron, but also a new particle, the neutrino (mass of the order of the electronic mass or smaller; no charge) is emitted."]

energy of the system: (1) the energy contribution due to the heavy particles present in the nucleus—protons and neutrons; (2) the contribution due to the light particles emitted or absorbed; and (3) the contribution due to interaction between light and heavy particles.

There are strong similarities between Fermi's account of nuclear structure here and Heisenberg's. But the comparison is not exact. (Fermi's method employs the so-called "second quantization" approach of Wigner and Jordan and this means that his mathematical expressions for dynamically describing the system of particles are of an entirely different order from Heisenberg's.)

In spite of the difference in formalism (better, because of it) Fermi succeeds in giving an adequate account of the beta-decay problem. He is able to deduce a probability distribution for the occurrence of a neutron-proton transformation and hence for the occurrence of the beta-emission. And his results are in reasonable agreement with the observed facts. Indeed, "one can say that the comparison of theory and experiment gives as good an agreement as could be expected." [15] In addition, it also becomes possible to specify fairly definitely the conditions under which beta-emission can and cannot occur.

The remarkable success of Fermi's theory of beta-emission was short-lived. Almost at once notice of a difficulty was taken. The Russian physicist, I. Tamm, who with his colleague and countryman D. Iwanenko had been independently exploring the possibility of accounting for beta-

---

[15] *Ibid.*, p. 176: "Zusammenfassend kann man sagen, dass dieser Vergleich von Theorie und Erfahrung eine so gut Ubereinstimmung gibt, wie man nur erwarten konnte."

emission on the Heisenberg hypothesis, sought to compute the binding energy for the atomic nucleus implied by Fermi's theoretical formulation.[16] The result was disappointing in the extreme. According to Tamm's calculations the exchange energy between a neutron and proton would be given by the expression:

$$A(r) = \pm \frac{g^2}{16\pi^3 c \hbar r^5} I(r).$$

Here g is a constant determined by the Fermi theory, r is the distance between proton and neutron, c is the velocity of light, $\hbar$ is Planck's constant divided by $2\pi i$ and $I(r)$ is a decreasing function of r which is equal to 1 when $r = \hbar/mc$ (m being the mass of an electron). This implies, says Tamm, that the absolute value of $A(r)$ is less than 10^{-85} r^{-5} erg. He concludes:

> Thus $A(r)$ is far too small to account for the known interaction of neutrons and protons at distances of the order of $r = 10^{-13}$ cm. . . . Our negative result indicates that either the Fermi theory needs substantial modification (no simple one seems to alter the results materially), or that the origin of the forces between neutrons and protons does not lie, as would appear from the original suggestion of Heisenberg, in their transmutations, considered in detail by Fermi.[17]

---

[16] I. Tamm, "Exchange Forces between Neutrons and Protons and Fermi's Theory," *Nature,* Vol. 133 (June 30, 1934), p. 981; D. Iwanenko, "Interactions of Neutrons and Protons," *ibid.*

[17] Tamm, *op. cit.*

Fermi's theory had overcome the beta-emission problem and saved the principle of conservation of energy. But it led straight back to the persistent threat that atomic nuclei are dynamically unstable in the extreme—too unstable to exist.

The Fermi theory, as we said earlier, is a good "first approximation" to a solution of the beta-decay problem. It is a better approximation than the Heisenberg treatment, both in terms of its capacity to predict what measurements should reveal and in its capacity to clear up puzzles threatening the bases of quantum mechanics. It is important to note that in neither case is the quantum theory proper tampered with. The Schrödinger equation (in relativistic form) and the principle of conservation of energy continue to be taken as a contextual background for all of the reasoning that goes on. Logically, it is all very reminiscent of Newton's attempts to locate the cause of the spreading spectrum or Leverrier's search for an adequate explanation of the behavior of Uranus and Mercury. Heisenberg and Fermi are groping for the proper quantum-mechanical description of nuclear structure on the assumption that it is the description of the nucleus—not fundamental mechanical theory—which is at fault. Fundamental theory provides the vital clues which direct the inquiry, just as it did for Leverrier and Newton. The explanatory resources of the theory are probed thoroughly before anyone begins to talk about "defeat."

The logic of the Heisenberg-Fermi investigations amply illustrates what was said in Chapter 3 concerning "Approximative Inference" and what was said in Chapter 4 about the a priori in statistical theories. Heisenberg uses the background of quantum theory to draw approximative inferences about binding energies and beta-decay. Fermi improves the accuracy of the predictions using essentially the same theoretical framework (relativistically modified, of course). Both leave loose ends lying about. Neither hits the data on the nose.

THE YUKAWA PARTICLE

The anomaly was by this time well-defined. The Fermi account of beta-emission—compatible with the general quantum theory and seemingly confirmed by its success in predicting the continuous distribution of emission velocities—implied a binding energy between neutron and proton which is incompatible with energies found in experimental data. In fact, the energy would be too weak to hold the nuclei of atoms together at all: we should be led to expect the immediate dissolution of all physical objects. In a seemingly distant field, the so-called "hard" or penetrating component of cosmic radiation was about to raise its ugly head. And in Japan, Hideki Yukawa was engaged in work which would weave both of these strands into a coherent cloth.

Yukawa's paper was presented November 17, 1934, less than six months after the appearance of the letters by Tamm and Iwanenko announcing difficulty with the Fermi theory. His central hypothesis is couched in the following language:

To remove this defect [noted by Tamm and Iwanenko], it seems natural to modify the theory of Heisenberg and Fermi in the following way. The transition of a heavy particle from neutron state to proton state is not always accompanied by the emission of light particles, i.e., a neutrino and an electron, but the energy liberated by the transition is taken up sometimes by another heavy particle, which in turn will be transformed from proton state into neutron state. If the probability of occurrence of the latter process is much larger than that of the former, the interaction between the neutron and the proton will be much larger than in the case of Fermi, whereas the probability of emission of light particles is not affected essentially.[18]

In view of the tendency of some present day textbook writers to dismiss Yukawa's Nobel Prize work with a casual remark to the effect that "he predicted the existence of a previously undetected elementary particle" it is well to stress the deep complexity of this hypothesis, its strong originality and its amazing power to unify disparate and seemingly unconnected regions of nuclear physics.

What Yukawa is assuming is that the binding energy of the nucleus is supplied not by Heisenberg's electron-field, nor by Fermi's electron-neutrino field, but by a stronger field engendered by the electron-neutrino energy plus the energy of a new, medium-weight particle. He proposes to redescribe the makeup of the nucleus completely! Unlike the electron and neutrino, this new particle would seldom be emitted from the nucleus. Normally it would manifest itself only as a strong field over the short intranuclear distances. Two processes would therefore be available for the

---

[18] Hideki Yukawa, "On the Interaction of Elementary Particles," *Proceedings of the Physico-Mathematical Society of Japan,* Vol. 17 (1935), p. 48; reprinted in *Foundations of Nuclear Physics,* ed. R. T. Beyer (New York: Dover Publications, 1949), p. 139.

transformation of a neutron into a proton: (1) the transformation could be effected by the simultaneous emission of a neutrino and electron or (2) the transformation could be effected when the "Yukawa particle" is "exchanged" between a proton and neutron, changing each into the other type of particle. The latter process would be far more probable than the former. Hence, unlike the Fermi account, the Yukawa version depicts the nucleus as the locus of almost continual interchange of charge between the nucleons. The shift does not merely occur when beta-emission is experienced, it happens all of the time!

The formal development of this hypothesis involves several points of interest. For one thing, Yukawa employs a fertile analogy between the electromagnetic field which gives rise to the emission of a photon from an atom and the special field corresponding to his hypothesized particle. The well-known wave equation which characterizes the former field is

$$\left\{ \nabla^2 - \frac{1}{c^2} \frac{\partial^2}{\partial t^2} \right\} U = 0. \tag{A}$$

This equation has the solution $U = 1/r$ which corresponds to the Coulomb potential between an electron and proton located at a distance r from one another.

Now experimental data on scattering of protons and neutrons had shown that the intranuclear binding forces must be somewhat different in character from Coulomb forces. Not only must they be stronger at short distances, they must also drop off rapidly at moderately large distances. What is needed, says Yukawa, is a value of U in the neighborhood of $\pm g^2 \dfrac{e^{-\lambda r}}{r}$, where g and λ are constants of the appropriate dimensions. A potential of this form would clearly drop off more rapidly than the simple $1/r$ of the Coulomb forces but could be made very large at short distances by proper selection of g and λ. The question is: what sort of wave equation leads to a solution of this type? Yukawa finds as the simplest answer to this question the wave equation

$$\left\{ \nabla^2 - \frac{1}{c^2}\frac{\partial^2}{\partial t^2} - \lambda^2 \right\} U = 0. \tag{B}$$

Thus, except for the extra term in λ^2, the field characterizing Yukawa's particle is just like the electromagnetic field.

The remarkable fact about this choice of equation as a description of the intranuclear field is that it leads directly back to the formal treatments of binding energy by Fermi and Heisenberg. Heisenberg's "exchange force integral" $J(r)$—mentioned earlier—can be taken as Yukawa's $-g^2\dfrac{e^{-\lambda r}}{r}$ and the entire Heisenberg theory of beta-decay phenomena can be derived from Yukawa's equations! Heisenberg's account of intranuclear binding forces is therefore preserved almost to the letter. Indeed, it is completed by the above choice of function for $J(r)$ and brought even more closely into line with experimental data.

A similar link with the Fermi beta-emission theory is established by Yukawa in the latter part of his paper. There he shows that the probability distribution for beta-decay predicted by Fermi continues to hold even if the new particle is admitted into the nuclear family. Again the Yukawa hypothesis is shown to accord with the experimental facts to precisely the same degree as existing theory. And yet it is capable of resolving difficulties about binding energies which had plagued the Fermi account.

In addition to spinning out connecting threads to the

Fermi and Heisenberg formalisms, Yukawa further elaborates his basic hypothesis. Like Leverrier almost a century before he is positing the existence of an object as yet unknown. It is likewise incumbent upon him to say what properties the object can be expected to have and also to explain why it has not previously been detected. The most essential property, of course, is the rest mass of the particle. But almost equally important are its charge and its tendency to be emitted from the nucleus.

On the matter of charge Yukawa observes that "the law of conservation of the electric charge demands that the quantum should have the charge either $+e$ or $-e$." [19] Determination of the particle's mass is not so direct a matter, however. The details of Yukawa's reasoning are discussed in Appendix B.[20] The upshot of the slightly elaborate argument is that the new particle must have a mass some 200 times as great as the mass of an ordinary electron or positron. The particle must be a middle-weight object somewhere between protonic and electronic size.

Concerning the role of the new particle in the nucleus, Yukawa shows from experimental data that the value of the constant g—which indicates the so-called "mesic charge" analogous to the electric charge e of an eletcron—is rather large. In fact it is about three times the value of the electronic charge. Hence the binding force between a neutron and proton due to their "sharing" a Yukawa particle is

---

[19] *Ibid.,* p. 52 [143]. The bracketed page number is the pagination of the reprint.

[20] See Alan M. Thorndike, *Mesons, A Summary of Experimental Facts* (New York: McGraw-Hill Book Co., 1952), pp. 20–23. Our discussion closely follows that of Thorndike.

some three times greater than the repulsive force of two protons. The Yukawa particle is more than enough to keep the nucleus intact.

Finally, Yukawa's calculations show that the conditions under which his particle can be emitted from the nucleus are extremely improbable in nature. If W_N denotes a neutron's energy state and W_P the energy state of a proton, then the condition under which the neutron and proton can give rise to emission of a particle of mass m is that $|W_N - W_P| > mc^2$. "The reason why such massive quanta, if they ever exist, are not yet discovered may be ascribed to the fact that the mass m is so large that [the condition] is not fulfilled in ordinary nuclear transformation." [21] In general, the energy states of nucleonic neutrons and protons just do not differ that much.

With this description of the properties of the Yukawa particle it would be only a matter of time before the basic hypothesis would either be confirmed or falsified. Experimental physicists were provided with the "vital statistics" -–mass 200 m_e, strongly interacting with nucleons, bearing either positive or negative electric charge—and it was up to them to apprehend the culprit if it showed its face. Yukawa even suggested where to look. (Shades of Leverrier!)

> Such quanta, if they ever exist and approach the matter [sic] close enough to be absorbed, will deliver their charge and energy to the latter. If, then, the quanta with negative charge come out in excess, the matter will be charged to a negative potential. These argu-

---

[21] Yukawa, *op. cit.*, p. 54 [145].

ments, of course, of merely speculative character, agree with the view that the high speed positive particles in the cosmic rays are generated by the electrostatic field of the earth, which is charged to a negative potential. The massive quanta may also have some bearing on the shower produced by cosmic rays.[22]

The exact process Yukawa envisioned in this passage is not quite clear. Possibly he was suggesting that the earth's negatively charged electrostatic field was produced by capture of negative Yukawa particles in the upper atmosphere creating a field which would accelerate positive particles (Yukawa particles?) toward the Earth as cosmic radiation. In any case, the general area in which the Yukawa particle is to be sought comes through unambiguously. In Yukawa's view the particle has something to do with the production of the positive component in the cosmic rays.

How right Yukawa's suggestion was began to become clear as further data on the cosmic rays filtered in. First, the conjecture as to the photon-constitution of the original cosmic rays at the top of the atmosphere was vigorously challenged during and after the 1934 International Congress on Nuclear Physics at London. Results reported by a 1933 Dutch expedition to Java corroborated definitely that the primary radiation underwent a variation according to latitude and hence must be corpuscular. In March of 1935 the Dutch physicist, J. Clay,[23] called these results (and others dealing with observations in coal mines) to the atten-

---

[22] *Ibid.*, p. 57 [148].

[23] See J. Clay, "The Nature of Cosmic Rays," *Proceedings of the Royal Society of London A,* Vol. 151 (1935), pp. 202–10.

tion of Millikan, Bowen, Neher, Anderson, and Neddermeyer, Compton, and Regener—who had all assumed in discussions at the 1934 Congress that the primary radiation is primarily made up of photons.

Evidence also had begun to accumulate against the identification of the penetrating component with protons and the general consensus in 1935 was that these particles must be positrons. By 1936, however, even this identification had become dubious. For one thing, improved understanding of the shower-producing process showed that the showers were engendered by photons, electrons and positrons, the latter acting as "radiators" emitting photons whenever passing near nuclei. The photons, in turn, produced "pairs"—electron plus positron—to keep the process going. Hence the shower-producing component's behavior was quite normal quantum mechanically and its energy loss accorded exactly with the so-called Heitler-Bethe radiation law deducible from relativistic quantum mechanics. Bruno Rossi describes the conceptual shift this provoked:

The theory of radiation losses and pair production, along with the consequent explanation of showers, shifted the mystery from one component of the local [secondary] radiation to another. To recapitulate, before the theory had made possible any precise predictions of the behavior of high-energy electrons and photons, physicists had thought the penetrating group of particles were electrons, which in traversing matter lost energy mainly by ionization. Since they were unaware of the overwhelming predominance of radiation processes at high energies, they had only to postulate electron energies of a few BeV in order to explain the tremendous penetrating power of these particles. The great puzzle, then, was the nature of the so-called shower-producing radiation. Obviously, the production of showers involved something more than ionization losses by electrons and Compton collisions by photons. After learning about pair production and radiation processes, physicists realized that the shower-producing radiation consisted of high-energy electrons and photons behaving exactly as they ought to behave. But, at the same time, it was equally evident that the penetrating particles could not possibly be electrons behaving in accordance with the theoretical predictions.[24]

In short, the highly penetrating component appeared neither to consist of protons nor positrons. But what else was left? There were no other kinds of positively charged particles to be considered.

Then, on the basis of 10,000 cloud-chamber photos taken atop Pikes Peak and 10,000 or so more taken at Pasadena, Anderson and Neddermeyer in 1936 came to the following distressing conclusions:

> . . . the rapid increase in the number of showers with increasing altitudes indicates that the photons and electrons producing the showers are highly absorbable and have a high probability of secondary electron production. This large absorbability of electrons is difficult to reconcile with the highly penetrating character of a large fraction of the sea-level particles on the view that the latter are electrons. It is therefore important to identify the penetrating particles. Some difficulties with identifying a large fraction of them with protons have

---

[24] Rossi, op. cit., p. 101.

already been discussed by Bowen, Millikan and Neher, and by ourselves. Furthermore experiments that we have carried out at Pasadena designed to bring into evidence high energy primary particles of protonic mass, should they exist in appreciable numbers, have so far given negative results.[25]

The problem had by this time become clear and acute. Two factors were involved: (1) the mere taxonomy of the penetrating component was in doubt, and (2) the extreme intensity and excessive energy loss through ionization of the penetrating particles above 1,000 MEV was incompatible with the predicted high values for radiation plus ionization energy loss. Just what was at stake could be seen clearly in one of Anderson and Neddermeyer's published photographs of a strongly ionizing particle penetrating a lead plate and passing on. The caption reads:

Below the plate it shows a greater ionization than an electron, and is deviated in the magnetic field to indicate a positively charged particle. Its $H\rho$ is apparently at most 1.4×10^5 gauss cm, which corresponds to a proton energy of 1 MEV and a range of only 2 cm in the chamber, whereas the observed range is greater than 5 cm.[26]

The ionization energy loss of the particle in traversing that range with such strong ionization would clearly be

---

[25] S. H. Neddermeyer and C. D. Anderson, "Cloud Chamber Observations of Cosmic Rays at 4300 Meters Elevation and Near Sea-Level," *Physical Review*, Vol. 50 (1936), pp. 268–69.
[26] *Ibid.*, p. 270.

greater than the loss theoretically possible for either the
positron or the proton. Or, to put it the other way around,
the particle's apparent energy loss through radiation would
have to be far less than theory allowed.

The decisive fact here is the clear evidence presented by
the ionization. The particle cannot be an electron. Basic
theoretical commitments rule that out completely. So it
has to be a proton. But its range of penetration is outside
the maximum possible range of penetration for a 1 MEV
proton! More directly and immediately than any previous
data the Anderson-Neddermeyer photo calls for a reassess-
ment of the basic assumptions of the past. *The particle
in Anderson and Neddermeyer's photograph is behaving in
such a way as to challenge existing theory all by itself.*

Yet Anderson and Neddermeyer exercised appropriate
caution in the face of this evidence. After all, only 11
of the Pasadena photographs and 113 of the Pikes Peak pictures were unambiguously interpretable as showing heavily
ionizing particles of positive charge. Further investigation
was required. Another possibility had to be ruled out.
Anderson and Neddermeyer went only so far as to say
that *either* there is a difference in character among the
particles *or* the absorption laws vary with energy. Through
the lenses of 20–20 hindsight it is tempting for us to
wonder why Anderson and Neddermeyer did not regard
the evidence at hand as sufficient to warrant the immediate
assumption of a wholly new species of particle. A little
reflection, however, shows why the cautious approach was
mandatory. The known "breakdown" of the radiation laws
governing ionization of the penetrating particles had been
discovered in the very high energy range—1,000 MEV
particles and above. This could conceivably be explained
away by the assumption of a special radiation law applying

only at the high energy levels. There was, in other words, every reason to believe that the "breakdown" obtained systematically and uniformly at the high energy levels and only there. Particles like that of Anderson and Neddermeyer's photograph (a fairly low energy specimen) did not fit well into this picture. For they were obviously rarities. And quite clearly the violation of Heitler and Bethe's radiation law by high energy particles was not a rare occurrence. Very possibly the kind of things which showed up in Anderson and Neddermeyer's photographs represented only a defect in experimental procedure, not a new particle. And on account of the energy range in which they occurred, it was still possible that they could be protons. This very real possibility—*not Humian scepticism about all forms of generalization*—prompted Anderson and Neddermeyer to exercise restraint.[27] Before one could claim that there are particles at *all* energy levels acting in violation of the Heitler-Bethe law one needed more data.

What was in question at this stage in the inquiry was the *scope* of the Heitler-Bethe law. Does it hold only at low energies? To what extent is it violated at the higher energies? And so forth. All of these questions are straightforwardly empirical. They can be settled by experiment and observation. Unlike the more fundamental principles of quantum mechanics, therefore, the Heitler-Bethe law was not "functionally a priori" or invulnerable to piecemeal change. Still, if no explanation of the high energy breakdown and the odd particles of Anderson and Nedder-

---

[27] For a discussion of the deductive character of generalization, Humian scepticism and the problem of induction see Chapter 4 *supra*.

meyer turned up the foundations of the theory would crumble. For the scope of the basic principles surely could not be modified willy nilly to exclude all of the particles behaving anomalously.

While Anderson and Neddermeyer further pursued the topic of the heavily ionizing particles in the United States, P. M. S. Blackett and his associate J. G. Wilson in England and Crussard and Leprince-Ringuet in France were pressing onward with their investigation of essentially the same matter. Using specially designed and constructed equipment—a combination Wilson chamber-counter arrangement—to get better pictorial data on the rays, Blackett and Wilson probed the breakdown in the radiation law at high energies with an eye to determining whether it was "systematic" or not. *In the course of their investigation they encountered only three of the heavily ionizing particles mentioned by Anderson and Neddermeyer, and these were presumed to be low energy protons.* They did, however, discover that the radiation law failure was fairly systematic and bounded within a limited range seemingly centering in the neighborhood of 2.5×10^9 e-volts.

These results, which are shown to be roughly consistent with the previous results of Anderson and Neddermeyer up to energies of 4×10^8 e-volts and those of Crussard and Leprince-Ringuet, are in striking contrast with the predictions of quantum mechanics . . . There is no possibility of avoiding this discrepancy by assuming protons in the beam, as all the particles with energy less than 6×10^8 e-volts are recognizable as electrons.[28]

---

[28] P. M. S. Blackett and J. G. Wilson, "The Energy Loss of

In order to "patch up" the theory, Blackett and Wilson assumed (following Nordheim) that the standard radiation absorption formula is violated at high energies. The new radiation formula thus suggested, however, was in violation of the relativistic requirement of Lorentz invariance and hence could not be taken too seriously.

The nature of the problem as Blackett and Wilson had nailed it down is this:

Between $E = 0$ and $E = 3 \times 10^9$ e-volts the [observed] energy loss is between two and three times that due to ionization alone; between $E = 3 \times 10^9$ and 10^{10} e-volts the energy loss is but little greater than the ionization loss; for energies between 10^{10} and 2×10^{10} e-volts the energy loss again rises to over three times the ionization loss. The observed excess energy loss is attributed to loss by radiation emission and shower formation.[29]

All of this on the assumption that the particles are positive electrons. Thus, in the intermediate energy range (3,000 MEV to 10,000 MEV) the calculated energy loss due to radiation emission was far lower than was to be expected. The particles simply penetrated too far.

Cosmic Ray Particles in Metal Plates," *Proceedings of the Royal Society of London A,* Vol. 160 (1937), p. 322.

[29] P. M. S. Blackett, "The Energy-Range Relation for Cosmic Ray Particles," *Proceedings of the Royal Society of London A,* Vol. 159 (1937), p. 31. Note specifically the "cut-off" value of 3×10^9 electron volts. The assumption underlying this is that the low energy particles of high penetration which had been observed are not to be taken into account.

By the spring of 1937, Anderson and Neddermeyer finally convinced themselves that a new type of particle— or perhaps we should say a new *form* of particle—was responsible for the absorption and radiation anomalies that had been observed since 1934. Having confirmed their earlier observations and supplemented them with further data from the low to intermediate energy range, Anderson and Neddermeyer claimed to have established

. . . the first experimental evidence for the existence of particles of both penetrating and non-penetrating character in the energy range extending below 500 MEV. Moreover, the penetrating particles in this range do not ionize perceptibly more than the non-penetrating ones, and cannot therefore be assumed to be of protonic mass. The lowest H_ρ among the penetrating group is 4.5×10^5 gauss cm. A proton of this curvature would ionize at least 25 times as strongly as a fast electron.[30]

The penetrating particles, then, were too penetrating to be ordinary electrons and positrons and not sufficiently different from positrons in ionizing power to be protons.

The non-penetrating particles are readily interpreted as free positive and negative electrons. Interpretations of the penetrating ones encounter very great difficulties,

---

[30] S. H. Neddermeyer and C. D. Anderson, "Note on the Nature of Cosmic Ray Particles," *Physical Review*, Vol. 51 (1937), p. 886.

but at present appear to be limited to the following hypotheses: (a) that an electron (+ or −) can possess some property other than its charge and mass which is capable of accounting for the absence of large radiative losses in a heavy element; or (b) that there exist particles of unit charge, but with a mass (which may not have a unique value) larger than that of a normal free electron and much smaller than that of a proton; this assumption would also account for the absence of numerous large radiative losses, as well as for the observed ionization. Inasmuch as charge and mass are the only parameters which characterize the electron in the quantum theory, assumption (b) seems to be the better working hypothesis. If the penetrating particles are to be distinguished from free electrons by a greater mass, and since no evidence for their existence in ordinary matter obtains, it seems likely that there must exist some very effective process for removing them. The experimental fact that penetrating particles occur with both positive and negative charges suggests that they might be created in pairs by photons, and that they might be represented as higher mass states of ordinary electrons.[31]

It is extremely noteworthy that Anderson and Neddermeyer did *not* assume simply and straightforwardly that a new *species* of fundamental particles had been found. They were careful to leave open the possibility that the cosmic ray particles were *reducible* to known classes of particle via special assumptions about states of the latter. This is in sharp contrast to the understanding of their work some-

---

[31] *Ibid.*

times conveyed in the later physical literature by the phrase "discovery of a new fundamental particle."

Simultaneous (or nearly so) with Anderson and Neddermeyer's announcement came word from Street and Stevenson of further confirmation and supporting evidence to back up their conclusions.[32] Neddermeyer and Anderson's paper was received by the *Physical Review* on March 30 and a note in proof had to be added at the last moment in order to acknowledge Street and Stevenson's report to the American Physical Society's April 29 meeting. In summarizing their experimental findings, Street and Stevenson not only corroborate Anderson and Neddermeyer but they also call attention vigorously to the clearest evidence that Blackett's assumption of a breakdown in the absorption laws will not suffice. This evidence consists in the fact that the secondary particles produced in showers—and known to be electrons and positrons—do not exhibit the radiation and absorption anomalies which the penetrating particles show. If, as Blackett and Wilson had guessed in 1936, it was merely a matter of the radiation energy loss being variable with the energy of the particles, then the same effects ought to be observed among shower particles of the same or comparable energies. Such effects do not occur.

Even before the conclusions of Anderson, Neddermeyer, Street, and Stevenson reached England, the Blackett-Wilson "breakdown" conjecture had come under heavy fire. In a strongly-worded paper written at almost exactly the

---

[32] J. C. Street and E. C. Stevenson, "Penetrating Corpuscular Component of the Cosmic Radiation," *Physical Review*, Vol. 51 (1937), p. 1005. The note appears on p. 886 of Neddermeyer and Anderson's paper, *op. cit.*

same time as Blackett and Wilson's, the noted theoreticians
H. J. Bhabha and W. Heitler (the latter being the co-author
of the radiation-absorption law which supposedly "broke
down") laid down the line beyond which the quantum
theory could not be forced.

We may describe the theory which is put forth in this
paper as the normal quantum theory of showers, inasmuch as it depends only on describing the interaction
of matter and radiation by Dirac's relativistic wave
equation, and the quantum theory of radiation. The
limits of our theory are therefore the limits of relativistic quantum mechanics . . .

It is our aim to deduce results which can be compared
directly with cosmic ray experiments and which will
then allow one to decide whether or not the theory
fails for extremely high energies, and in the latter case,
at what point the failure begins.[33]

Subsequently it was to be found that the very point at
which the "failure" begins is so low as to imply not only
a breakdown of the radiation law but a more fundamental
collapse of the whole quantum theory. Thus, Heitler and
Bhabha's results already contain the seeds of the destruc-
tion of Blackett's "breakdown hypothesis." In April of 1937
Heitler submitted a paper to the Royal Society in which
the seeds can be seen in the process of germinating.[34] Citing

---

[33] H. J. Bhabha and W. Heitler, "The Passage of Fast Electrons and the Theory of Cosmic Showers," *Proceedings of the
Royal Society of London A*, Vol. 159 (1937), pp. 432–34.

[34] W. Heitler, "On the Analysis of Cosmic Rays," *Proceedings
of the Royal Society of London A*, Vol. 161 (1937), pp. 261–83.

the calculations he and Bhabha had carried out at the end
of the previous year he argued that the "breakdown hypothesis" plus empirical data on the intensity of the rays at
high altitudes is "incompatible with the assumption that
the breakdown point E_c lies appreciably lower than 3×10^9
e-volts." [35] In a note added in proof Heitler takes cogni-
zance of Blackett and Wilson's results leading to a breakdown point possibly as low as 2 or 3×10^8 e-volts and adds
a curious allusion to Neddermeyer and Anderson's paper as
follows:

> Finally we may mention that in another recent paper,
> Neddermeyer and Anderson (1937) pointed out a few
> arguments in favour of a new sort of particles with
> mass between an electron and proton . . .[36]

Heitler's coolness to the Anderson-Neddermeyer proposal
notwithstanding, his own results had served to lay to rest
once and for all the competing "breakdown hypothesis."
    Blackett soon came forward to recant and to join Nedder-
meyer and Anderson in affirming the existence of a new
species of particle-track: "We concluded [in our earlier
work] that the energy loss of a normal electron varies with
its energy. We now believe this to be probably false . . ." [37]
But Blackett did not go on to affirm the view that the
particles discovered constituted a new addition to the fam-

---

[35] *Ibid.,* p. 279. See note 29 above.
[36] *Ibid.,* p. 282.
[37] P. M. S. Blackett, "The Nature of the Penetrating Component of Cosmic Rays," *Proceedings of the Royal Society of
London A,* Vol. 165 (1938), p. 11.

ily of fundamental particles. Instead, he adopted Anderson and Neddermeyer's suggestion that the particle is an electron in a special state of "heavy mass."

> It is clear that the penetrating rays cannot be heavy electrons of constant rest mass, or of any mixture of such heavy electrons, since the mean radiation loss varies rapidly. They might, however, consist of heavy electrons with a variable rest mass, for instance, 'excited' electrons which go to their normal state when their energy drops below the critical value. Alternatively, the heavy particle may in some way give rise to, rather than become, the light particles.[38]

The problem which prompts Blackett to adopt this tentative viewpoint is a real one. For at very low energies the "heavy electrons" are no different from ordinary electrons in their energy losses and ionization. (Or, at least, this is the way it appears. In point of fact, the last alternative mentioned by Blackett is the one which obtains: the "heavy electrons" decay into ordinary positrons and electrons. But this was not immediately recognized.)

In all of the discussion so far considered no mention is made of Hideki Yukawa's conjecture on the existence of a new fundamental particle. The reason is simply that the option of regarding the absorption anomalies as stemming from variations in the mass of an ordinary electron could not be ruled out *a priori*. And within a short time another possibility had emerged: that the anomalies were produced not by one new particle but by a whole host of them each

---

[38] *Ibid.*, p. 26.

with a different mass! By the end of 1937, however, the Yukawa theory had been associated with the problem by a number of writers, including Oppenheimer and Serber, Stueckelberg, and Yukawa himself.[39] Work was begun to bring the theory into comparison with the observations on the cosmic ray particles. To all intents and purposes the Yukawa particle had been found. And when, in January of 1938, Blackett and Wilson conceded that "it does not seem likely that the present quantum theory is adequate to describe the supposed change of rest mass" involved in the "variable-mass" hypothesis, Yukawa's triumph seemed assured.

THE AFTERMATH

Yukawa's enjoyment of success was over almost before it began. Within a very few months evidence began to pile up implying that his theoretical predictions of the properties of the "heavy quantum" were not only very "qualitative"—i.e., approximative—but that they were grossly in error. Large-scale modifications of the theory were undertaken, which Yukawa later described in these terms:

. . . the [original] simple theory was incomplete in various respects. For one thing, the exchange force thus obtained was repulsive for triplet S-state of the deuteron in contradiction to the experiment, and moreover we

---

[39] For a list of references to the literature see N. Kemmer, "Einstein-Bose Particles and Nuclear Interaction," *Proceedings of the Royal Society of London A,* Vol. 166 (1938), p. 127 (footnote).

could not deduce the exchange force which was necessary in order to account for the saturation of nuclear forces just at the alpha-particle. In order to remove these defects, more general types of fields including vector, pseudoscalar and pseudovector fields in addition to the scalar fields, were considered by various authors.[40]

Generally, the pseudoscalar field was found to accord most nearly with the experimental data. All of the more general types of fields, however, gave rise to intractable mathematical problems due to divergences within the expressions used to describe the field.

These modifications, of course, were made simply to bring the original hypothesis into line with data which were available before the discovery of the "heavy electron" of Anderson and Neddermeyer. As data from cosmic ray observations were added the association of the Yukawa particle with the "heavy electron"—or "meson" as Bhabha dubbed it[41]—came into question. Almost none of their properties tallied. Anomalies cropped up on all sides. They would have to be explained. The most important discrepancy—reported in 1939 by Nordheim and Hebb[42]—was the

---

[40] The Nobel Foundation, *Nobel Lectures in Physics, 1942–1962* (Amsterdam-London-New York: Elsevier Publishing Co., 1964), p. 129. For a discussion of the various types of meson fields see Wolfgang Pauli, *Meson Theory of Nuclear Forces* (New York: Interscience Publishers, Inc., 1946).

[41] H. J. Bhabha, *Nature*, Vol. 143 (1939), p. 276. Other names included 'mesotron,' 'heavy electron' and 'yukon.'

[42] L. W. Nordheim and M. H. Hebb, "On the Production of the Hard Component of the Cosmic Radiation," *Physical Review*, Vol. 56 (1939), pp. 494–507.

fact that the cosmic ray meson's creation in the upper atmosphere required the assumption of an interaction cross-section far greater than a particle of such penetrability could have. This conclusion was independent of whether the mesons were created by the impact of photons or protons against nuclei in the upper atmosphere. And its implication (not stressed by Nordheim and Hebb) is that the binding energy of the nucleus is too large to be the result solely of the cosmic-ray particle's field.

Heitler and Ma in 1940 make it explicit:

The meson theory in its present form exhibits a number of serious difficulties if applied to the interaction of fast mesons with a nuclear particle. This interaction increases rapidly with increasing energy, and hence all quantities involving fast mesons diverge or are far too big to fit the experiments. The anomalous magnetic moment of the proton, for instance, diverges strongly [when calculated from the meson theory]. The cross-section for the scattering of fast mesons by nuclei is found [in calculations] to increase rapidly with energy which is contrary to the experiments, and even at small energies the cross-section is bigger by an order of magnitude than experiments permit.[43]

Heitler and Ma were moved by these considerations to propose a radical innovation in quantum theory: the assumption that nucleons have an additional "degree of

---

[43] W. Heitler and S. T. Ma, "Inner Excited States of the Proton and Neutron," *Proceedings of the Royal Society of London A*, Vol. 176 (1940), p. 369.

freedom" such that a particle of protonic mass would not be limited to spin $1/2$ and charge $\pm e$ but could have spins like $3/2$, $5/2$ . . . and charges of integral multiples of e, positive and negative. It is clear from the extreme character of this proposal just how seriously the difficulties in meson theory were being taken in 1940. Nor were Heitler and Ma the only writers proposing radical overhaul of the entire meson-Yukawa particle frame of reference. Bethe, motivated in part by the apparent equivalence of proton-proton, proton-neutron, and neutron-neutron attractions, was suggesting at about this time that the particle responsible for nuclear forces must be neutral, rather than positive or negative as assumed by Yukawa.[44] This approach led to fairly exact treatment of a narrow range of problems but, as Heitler and Ma observed in rejecting the Bethe hypothesis, ". . . in a preliminary theory like the meson theory . . . more weight should be put on a qualitative connected account of many different phenomena than on a quantitative treatment of a single effect."[45] Approximation is no blemish in the eyes of Heitler and Ma unless a better approximation can be found.

By 1941 the only justification for assuming the identity of the Yukawa particle and the cosmic-ray meson was, in Booth and Wilson's words, "that thereby the number of elementary particles required is kept as small as possible."[46] *Pace* Ockham. The anomalies with which the theory was

---

[44] H. Bethe, *Physical Review*, Vol. 57 (1940), pp. 260 and 390.

[45] Heitler and Ma, *op. cit.*, p. 384 footnote.

[46] F. Booth and A. H. Wilson, "Radiative Processes Involving Fast Mesons," *Proceedings of the Royal Society of London A*, Vol. 174 (1940), p. 515.

saddled included such formidable difficulties as the following:[47] (1) the only interaction of the Anderson-Neddermeyer particle with matter seemed to be due to its charge;
(2) scattering by interactions other than electric was small
and perhaps even non-existent; (3) the absorption of slow
mesons by nuclei took place with a probability far below
that expected; yet (4) a considerable fraction of mesons
produced in cosmic radiation penetrated all the way down
into the earth; and (5) they were produced in showers,
suggesting a cross-section larger than that exhibited in their
weak interactions with nuclei. And if this were not enough,
the observed lifetimes of the cosmic ray mesons were too
long to account for beta-decay.

With the start of World War II, active research into the
question was severely retarded and communications among
scientists of various nations were reduced to a minimum.
It was not until 1946, therefore, that a new attempt to
resolve the difficulties saw the light of public scrutiny
(though, according to Yukawa, the approach employed
had been invented in Japan as early as 1942).[48] The
Japanese physicists S. Sakata and Y. Tanikawa advanced
the conjecture[49] that there must indeed be two different

---

[47] See V. F. Weisskopf's letter to the editor, *Physical Review*,
Vol. 72 (1947), p. 510.

[48] See p. 131 of *Nobel Lectures* . . . , *op. cit.*, where Yukawa
says that the hypothesis was proposed in 1942. However, in
Yukawa's "Models and Methods in the Meson Theory," *Re-
views of Modern Physics*, Vol. 21 (1949), p. 475, the date is
given as 1943.

[49] S. Sakata and T. Inoue, *Progress in Theoretical Physics*, Vol.
1 (1946), p. 143; Y. Tanikawa, *Progress in Theoretical Physics*,
Vol. 2 (1947), p. 220.

middle-weight or mesonic particles; that the Yukawa particle and the particle discovered by Neddermeyer and Anderson are not one and the same entity. The former was assumed to be a heavier object and the one responsible for nuclear binding and beta-decay. Subsequently, the same proposal was made independently by Marshak at the Conference on the Foundations of Quantum Mechanics (June, 1947) at Shelter Island, N. Y. Marshak put the relative masses at 125 and 100 m_e.[50] Between the publication of Sakata's article and Marshak's talk, however, an important experimental finding intervenes which destroys once and for all the illusion that the Anderson-Neddermeyer cosmic ray particles could be capable of providing the immense forces required to hold the nucleus together. In Italy, Conversi, Pancini, and Piccioni discovered that negative cosmic-ray mesons captured by nuclei in a carbon plate frequently underwent a "delayed" decay into a negative electron (and a neutrino).[51] In fact, this "delayed" decay occurred more often in the carbon than in an iron plate. But according to Yukawa theory, the negative meson at rest is far more likely to be absorbed by a nucleus than to decay—and it is far more likely to be absorbed by the lighter carbon nuclei than by the iron nuclei. In a "delayed" decay the meson would nearly approximate the state of rest and should therefore be absorbed. That this regularly failed to

---

50 R. E. Marshak and H. A. Bethe, "Two-meson Hypothesis and the π-μ Decay Process," *Physical Review*, Vol. 72 (1947), p. 506.
51 Conversi, *et al.*, "On the Disintegration of Negative Mesons," *Physical Review*, Vol. 71 (1947), pp. 209–10. Letter to the editor of Dec. 21, 1946.

occur was clear proof that the cosmic-ray particle could not be Yukawa's particle.

The explanation of the results of Conversi and his colleagues was shortly forthcoming.[52] Marshak and Bethe showed that on the assumption of a heavy meson with spin 1/2 and a light meson with spin 0 the results could be rationalized. The light meson's infrequent absorption, they pointed out, would follow from the fact that the absorption process was not direct but involved an intermediate step. Given the light meson and a proton, the intermediate step would consist of the formation of three particles: (a) a heavy meson, (b) a neutrino, and (c) the proton. These three, in turn, would yield a nucleonic neutron. In order for this to occur, however, a second nucleon would have to be on hand to "carry off" an excess of momentum generated in the transformation. Hence the general improbability of the entire process in light nuclei.

Now it was a matter of finding the second meson somewhere. Employing newly developed photoemulsion techniques, Lattes, Muirhead, Occhialini, and Powell[53] soon succeeded in doing just that. Their photographic plates of high-altitude cosmic ray bombardments revealed the existence of a process in which a single cosmic-ray meson of the Anderson-Neddermeyer type was produced by the "decay" of a slightly different kind of middle-weight particle. The two particles appeared to differ slightly in mass but, so far as the experimenters could see, not by more

---

[52] Marshak and Bethe, *op. cit.*

[53] C. M. G. Lattes, H. Muirhead, G. P. S. Occhialini, and C. F. Powell, "Processes Involving Charged Mesons," *Nature,* Vol. 159 (1947), pp. 694–97.

than about 60 electron masses. Interestingly enough, the Lattes-Muirhead group was not at the time acquainted with Marshak's conjecture and those of Sakata and Tanikawa. They knew of no "two-meson" theory involving the decay of one type into the other. Still, they recognized immediately that the "heavy" particle at hand must have some bearing on the problem pointed up by the "Rome group."

When all of the data were in there was no reasonable doubt left about the matter: the "heavy" meson was indeed the particle predicted by Yukawa more than 12 years earlier. It had the appropriate life-time, the strong interaction characteristics, the spin, the charge—everything. In time, the new particle came to be known as the 'pion' and its decay-product—the Anderson-Neddermeyer meson —acquired the label of 'muon' or 'mu meson.' Deep theoretical problems, of course, remained. As the history of science shows, no scientific discovery—however illuminating—ever completely clears up all of the anomalies in the field. At least, not permanently. But Yukawa's conjecture had been far more illuminating than most and had provided grounds for experimental and theoretical inquiries which are even today being carried forward at a quickening pace.

The role of anomalies in the stimulation of research in the early meson theory is unmistakable and offers further confirmation of the philosophical thesis that anomalies govern the logic of explanation. Starting with the problems about beta-decay, nuclear binding, and cosmic rays Yukawa framed an explanatory hypothesis designed to dispel all perplexity. When further anomalies threatened this hypothesis, deeper investigations were launched. *The H–D model of theories notwithstanding, this is the normal course of scientific progress.* The discovery of the pion is a triumph for the faith Yukawa and his colleagues had in the basic quantum theory and the meson hypothesis. Had they followed the logical guidelines of the H–D model they would surely have abandoned the entire enterprise long before 1947.

Two years after the discovery of the pion by Lattes, *et al.*, Yukawa was awarded the Nobel Prize in Physics "For having predicted, as a result of his theoretical work on nuclear forces, the existence of mesons." [54] The following year, C. F. Powell—who had participated with Lattes in the discovery of the pion—was the recipient of the prize for his important work in cosmic ray research and his efforts in improvement of techniques of particle observation. Powell concluded his acceptance address with a remark which well sums up the ultimate influence of Yukawa's work and the work of the dozens of men who played a role in the development of the early meson theory:

> . . . the study of what might, in the early days, have been regarded as a trivial phenomenon has, in fact, led us to the discovery of many new forms of matter and many new processes of fundamental physical importance. It has contributed to the development of a picture of the material universe as a system in a state of perpetual change and flux; a picture which stands in great contrast to that of our predecessors with their fixed and eternal atoms.

---

[54] Niels H. deV. Heathcote, *Nobel Prize Winners in Physics, 1901–1950* (New York: Henry Schumann, 1953), p. 446.

And yet,

> We are only at the beginning of our penetration into what appears to be a rich field of discovery. Already . . . it seems that our present theoretical approach has been limited by lack of essential information; and that the world of the mesons is far more complex than has hitherto been visualized in the most brilliant theoretical speculations.[55]

---

[55] Quoted by Heathcote, *ibid.*, p. 460.

7

Epilogue

In the dynamic growth of scientific knowledge, theories play two distinct but interlocking roles: they explain nature's puzzles for us and they tell us which states of affairs are genuinely puzzling. The statistical or non-statistical character of the theory makes no real difference in this respect. What *does* matter is the approximative character of the laws of the theory. For in order both to identify and overcome anomalies, a theory must be able to "roll with the punches." It must possess rich explanatory resources to fall back on in time of crisis. At the same time, it must normally be capable of functioning without all of these resources being brought to bear in every instance.

The quantum theory, as we have seen, has such resources. In the development of the meson concept the theory served repeatedly to locate and explain anomalies. Yet the theory itself was in no substantial way altered by this. The confrontation with anomalies did not generate a major crisis (though at many points it threatened to). This, as the historian, T. S. Kuhn, has rightly contended, is how science normally proceeds.[1] The key to our logical understanding

---

[1] Kuhn, *The Structure of Scientific Revolutions, op. cit.,* Chapters I–V.

of the process is the cluster of ideas associated with scope and approximation. Heisenberg's description of the nuclear state in beta decay is a fair approximation; Fermi's is better; Yukawa's is better still. Yet the scope of the Schrödinger equation is variable enough to accommodate any of the three (in non-relativistic form). In other words, a full list of all energy contributions whatever is not needed in order to generate meaningful predictions about the behavior of atomic systems. Progress toward a correct understanding of the anomaly can be made without alteration of the basic postulates of quantum mechanics. This is not—as some logicians and H–D theorists might contend—a defect or flaw in the theory. *It is the theory's main source of explanatory power!* The fact that Yukawa's meson formalism is highly "qualitative"—i.e., leads to erroneous predictions— does not lead to the overthrow of the quantum theory or even Yukawa's own hypothesis. It merely signifies the need for further empirical and theoretical investigation of the anomalies. As this "looseness of fit" is tightened by the acquisition of new knowledge, the quantum theory can reasonably be expected to pile up more and more explanatory successes. Were the theory truly an axiomatic hypothetico-deductive system this would be out of the question.

We must not be misled, however, into believing that physical theories are completely "open-ended." As Bhabha and Heitler showed, there are limits beyond which quantum theory cannot be pushed. (This is the kernel of truth in the anti-Copenhagen arguments of David Bohm.) The quantum theory *can* be falsified if those limits are transgressed. Bhabha and Heitler explicitly recognized as much when they sought

. . . to deduce results which can be compared directly
with cosmic ray experiments and which will then allow
one to decide whether or not the theory fails for extremely high energies . . .[2]

The task of "drawing the line"—a task carried out for
Cartesian optics by Newton; for classical celestial mechanics by Leverrier; for Maxwellian electromagnetic theory by Einstein, Millikan and Marx—is a most important
one indeed. It is the method by which, ultimately, the
semantical content of a physical theory is laid bare.

What holds for the quantum theory holds equally for
other systems of physical explanation. Their explanatory
force and their utility arise directly from their approximative character—from their capacity to draw inferences from
less-than-complete information and to resolve anomalies as
they arise. These are the features which make a good theory
such a valuable epistemic tool for probing the ontological
structure of nature.

As an anomaly-identifier, a physical theory serves to provide a context within which important observations can
be sifted out from the unimportant. As an anomaly-explainer, the theory provides insight, understanding, and
illumination of those observations. To view a theory ex-
clusively in terms of one or the other of these functions—
in particular, to stress the explanatory aspect as the H–D
theorists do—is to misrepresent the logic of physical explanation. For physical explanation is always bound to

---

[2] "The Passage of Fast Electrons and the Theory of Cosmic
Showers," *op. cit.,* p. 433.

some conceptual and theoretical context. It is never carried on *in vacuo*. Mere deduction of consequences from hypotheses does not, in itself, constitute explanation. Only when set against a background—against an anomaly context—does the deductive inference provide genuine understanding. For this reason, if for no other, we should prefer to regard physical theories as systems of approximative generalizations rather than rigid axiomatic schemata.

Appendix A: The Bohm Reconstruction

In his 1952 paper, David Bohm considers the situation ordinarily treated in quantum mechanics by means of the one-particle Schrödinger equation. We can think of this as an analysis of the behavior of an electron in a field with potential $V(x)$. Now Bohm associates with this electron (a) a field of force characterized by something he calls a "quantum mechanical potential" and (b) a particle "having precisely definable and continuously varying values of position and momentum." Mathematically, this program is carried out as follows:

The one-particle Schrödinger equation can be written as

$$i\hbar \frac{\partial \psi}{\partial t} = \frac{(i\hbar)^2}{2m} \nabla^2 \psi + V(x)\psi. \tag{1}$$

Bohm here introduces two arbitrary real functions of x, R and S, and proposes that we regard the solution of the Schrödinger equation for ψ as expressed in terms of R and S; i.e., he assumes

$$\psi = R \exp \frac{iS}{\hbar}. \tag{2}$$

Differentiating equation (2) with respect to time and substituting into (1) we find that

$$\frac{\partial R}{\partial t} = -\frac{1}{2m} [R\nabla^2 S + 2\nabla R \cdot \nabla S] \tag{3}$$

$$\frac{\partial S}{\partial t} = -\left[\frac{(\nabla S)^2}{2m} + V(x) - \frac{\hbar^2}{2m} \frac{\nabla^2 R}{R} \right]. \tag{4}$$

Setting $P(x) = R^2(x)$—this quantity being the probability distribution $|\psi(x)|^2$ as can be seen from (2) above—and substituting into (3) and (4) we obtain

$$\frac{\partial P}{\partial t} + \nabla \left(P \frac{\nabla S}{m} \right) = 0 \tag{5}$$

$$\frac{\partial S}{\partial t} + \frac{(\nabla S)^2}{2m} + V(x) - \frac{\hbar^2}{4m} \left[\frac{\nabla^2 P}{P} - \frac{1}{2} \frac{(\nabla P)^2}{P^2} \right] = 0. \tag{6}$$

The last term of this latter differential equation, viz.

$$-\frac{\hbar^2}{4m} \left[\frac{\nabla^2 P}{P} - \frac{1}{2} \frac{(\nabla P)^2}{P^2} \right],$$

is defined by Bohm to be $U(x)$, the quantum-mechanical potential acting on the particle. Further, it is assumed that the velocity of the particle is given by the formula

$$v(x) = \frac{\nabla S(x)}{m}. \tag{7}$$

Utilizing this equation and either equation (4) or equation (6) above it is possible to obtain a solution for the particle's trajectory in terms of its initial position and momentum, the potential $V(x)$ and the potential $U(x)$. The equation of motion as given by Bohm is

$$m \frac{d^2 x}{dt^2} = -\nabla \left[V(x) - \frac{\hbar^2}{2m} \frac{\nabla^2 R}{R} \right].$$

Thus, if we could determine the initial conditions, exact predictions about the future locations and momenta of the particle could be made.

But there is a catch. As Bohm himself is quick to point out, the behavior of the quantum force whose potential is $U(x)$—the ψ-field as it is occasionally called—becomes uncontrollable when a measuring instrument is introduced.

It follows that the most information we can obtain about the future positions of the particle is given by the probability distribution $|\psi(x)|^2 = P(x)$: in an aggregate of such particles varying probabilities can be assigned to each of the possible positions in which the particles might be found. This is essentially the same procedure as is carried out for the electron proper in the usual interpretation of the quantum theory.

Appendix B: Yukawa's Prediction of Mesonic Mass

Yukawa's reasoning in deriving the mass of his predicted particle can be schematized in the following fashion.

The wave equation (B) describing the mesonic field can be rewritten in the form

$$\frac{\partial^2 U}{\partial t^2} = \left[c^2 \left(\frac{\partial^2}{\partial x^2} + \frac{\partial^2}{\partial y^2} + \frac{\partial^2}{\partial z^2} \right) - \lambda^2 c^2 \right] U \qquad \text{(B*)}$$

which is the Schrödinger equation of a freely moving particle. We now substitute the formal operators

$$\begin{cases} P_x = \dfrac{h}{2\pi i} \dfrac{\partial}{\partial x} \\[2mm] P_z = \dfrac{h}{2\pi i} \dfrac{\partial}{\partial z} \end{cases} \qquad \begin{cases} P_y = \dfrac{h}{2\pi i} \dfrac{\partial}{\partial y} \\[2mm] W = -\dfrac{h}{2\pi i} \dfrac{\partial}{\partial t} \end{cases}$$

into (B*) to obtain

$$-W^2 \left(\frac{2\pi}{h} \right)^2 = -c^2 \left(\frac{2\pi}{h} \right)^2 (P_x^2 + P_y^2 + P_z^2) - \lambda^2 c^2$$

which simplifies easily to

$$W^2 = c^2 P^2 + \frac{\lambda^2 h^2}{4\pi^2} c^2. \qquad \text{(B**)}$$

But the relationship of the momentum and energy of a particle of mass m is given in relativity theory by

$$W^2 = (cP)^2 + (mc^2)^2. \qquad \text{(C)}$$

Together (B**) and (C) entail that $m = \dfrac{\lambda h}{2\pi c}$. If $1/\lambda$ is approximately 2×10^{-13}—as scattering experiments seemed

307

to imply—then m, the mass of the new particle, would have to be approximately 200 times the mass of an electron. In other words, it would have to be intermediate between the masses of a heavy particle and a light one.

Index

Index